SOUND AND LIGHT

Anthea Maton
Former NSTA National Coordinator
Project Scope, Sequence, Coordination
Washington, DC

Jean Hopkins
Science Instructor and Department Chairperson
John H. Wood Middle School
San Antonio, Texas

Susan Johnson
Professor of Biology
Ball State University
Muncie, Indiana

David LaHart
Senior Instructor
Florida Solar Energy Center
Cape Canaveral, Florida

Maryanna Quon Warner
Science Instructor
Del Dios Middle School
Escondido, California

Jill D. Wright
Professor of Science Education
Director of International Field Programs
University of Pittsburgh
Pittsburgh, Pennsylvania

 Prentice Hall
Englewood Cliffs, New Jersey
Needham, Massachusetts

Prentice Hall Science
Sound and Light

Student Text and Annotated Teacher's Edition
Laboratory Manual
Teacher's Resource Package
Teacher's Desk Reference
Computer Test Bank
Teaching Transparencies
Product Testing Activities
Computer Courseware
Video and Interactive Video

The illustration on the cover, rendered by David Schleinkofer, shows that sound and light have both beautiful and practical applications which span the globe.

Credits begin on page 159.

SECOND EDITION

© 1994, 1993 by Prentice-Hall, Inc., Englewood Cliffs, New Jersey 07632. All rights reserved. No part of this book may be reproduced in any form or by any means without permission in writing from the publisher. Printed in the United States of America.

ISBN 0-13-400565-1

16   99

Prentice Hall
A Division of Simon & Schuster
Englewood Cliffs, New Jersey 07632

STAFF CREDITS

Editorial:	Harry Bakalian, Pamela E. Hirschfeld, Maureen Grassi, Robert P. Letendre, Elisa Mui Eiger, Lorraine Smith-Phelan, Christine A. Caputo
Design:	AnnMarie Roselli, Carmela Pereira, Susan Walrath, Leslie Osher, Art Soares
Production:	Suse F. Bell, Joan McCulley, Elizabeth Torjussen, Christina Burghard
Photo Research:	Libby Forsyth, Emily Rose, Martha Conway
Publishing Technology:	Andrew Grey Bommarito, Deborah Jones, Monduane Harris, Michael Colucci, Gregory Myers, Cleasta Wilburn
Marketing:	Andrew Socha, Victoria Willows
Pre-Press Production:	Laura Sanderson, Kathryn Dix, Denise Herckenrath
Manufacturing:	Rhett Conklin, Gertrude Szyferblatt

Consultants

Kathy French	National Science Consultant
Jeannie Dennard	National Science Consultant
Brenda Underwood	National Science Consultant
Janelle Conarton	National Science Consultant

Contributing Writers

Linda Densman
Science Instructor
Hurst, TX

Linda Grant
Former Science Instructor
Weatherford, TX

Heather Hirschfeld
Science Writer
Durham, NC

Marcia Mungenast
Science Writer
Upper Montclair, NJ

Michael Ross
Science Writer
New York City, NY

Content Reviewers

Dan Anthony
Science Mentor
Rialto, CA

John Barrow
Science Instructor
Pomona, CA

Leslie Bettencourt
Science Instructor
Harrisville, RI

Carol Bishop
Science Instructor
Palm Desert, CA

Dan Bohan
Science Instructor
Palm Desert, CA

Steve M. Carlson
Science Instructor
Milwaukie, OR

Larry Flammer
Science Instructor
San Jose, CA

Steve Ferguson
Science Instructor
Lee's Summit, MO

Robin Lee Harris
Freedman
Science Instructor
Fort Bragg, CA

Edith H. Gladden
Former Science Instructor
Philadelphia, PA

Vernita Marie Graves
Science Instructor
Tenafly, NJ

Jack Grube
Science Instructor
San Jose, CA

Emiel Hamberlin
Science Instructor
Chicago, IL

Dwight Kertzman
Science Instructor
Tulsa, OK

Judy Kirschbaum
Science/Computer Instructor
Tenafly, NJ

Kenneth L. Krause
Science Instructor
Milwaukie, OR

Ernest W. Kuehl, Jr.
Science Instructor
Bayside, NY

Mary Grace Lopez
Science Instructor
Corpus Christi, TX

Warren Maggard
Science Instructor
PeWee Valley, KY

Della M. McCaughan
Science Instructor
Biloxi, MS

Stanley J. Mulak
Former Science Instructor
Jensen Beach, FL

Richard Myers
Science Instructor
Portland, OR

Carol Nathanson
Science Mentor
Riverside, CA

Sylvia Neivert
Former Science Instructor
San Diego, CA

Jarvis VNC Pahl
Science Instructor
Rialto, CA

Arlene Sackman
Science Instructor
Tulare, CA

Christine Schumacher
Science Instructor
Pikesville, MD

Suzanne Steinke
Science Instructor
Towson, MD

Len Svinth
Science Instructor/
Chairperson
Petaluma, CA

Elaine M. Tadros
Science Instructor
Palm Desert, CA

Joyce K. Walsh
Science Instructor
Midlothian, VA

Steve Weinberg
Science Instructor
West Hartford, CT

Charlene West, PhD
Director of Curriculum
Rialto, CA

John Westwater
Science Instructor
Medford, MA

Glenna Wilkoff
Science Instructor
Chesterfield, OH

Edee Norman Wiziecki
Science Instructor
Urbana, IL

Teacher Advisory Panel

Beverly Brown
Science Instructor
Livonia, MI

James Burg
Science Instructor
Cincinnati, OH

Karen M. Cannon
Science Instructor
San Diego, CA

John Eby
Science Instructor
Richmond, CA

Elsie M. Jones
Science Instructor
Marietta, GA

Michael Pierre
McKereghan
Science Instructor
Denver, CO

Donald C. Pace, Sr.
Science Instructor
Reisterstown, MD

Carlos Francisco Sainz
Science Instructor
National City, CA

William Reed
Science Instructor
Indianapolis, IN

Multicultural Consultant

Steven J. Rakow
Associate Professor
University of Houston—
Clear Lake
Houston, TX

English as a Second Language (ESL) Consultants

Jaime Morales
Bilingual Coordinator
Huntington Park, CA

Pat Hollis Smith
Former ESL Instructor
Beaumont, TX

Reading Consultant

Larry Swinburne
Director
Swinburne Readability
Laboratory

CONTENTS
SOUND AND LIGHT

SCIENCE GAZETTES

Activity Bank/Reference Section

Features

CONCEPT MAPPING

Throughout your study of science, you will learn a variety of terms, facts, figures, and concepts. Each new topic you encounter will provide its own collection of words and ideas—which, at times, you may think seem endless. But each of the ideas within a particular topic is related in some way to the others. No concept in science is isolated. Thus it will help you to understand the topic if you see the whole picture; that is, the interconnectedness of all the individual terms and ideas. This is a much more effective and satisfying way of learning than memorizing separate facts.

Actually, this should be a rather familiar process for you. Although you may not think about it in this way, you analyze many of the elements in your daily life by looking for relationships or connections. For example, when you look at a collection of flowers, you may divide them into groups: roses, carnations, and daisies. You may then associate colors with these flowers: red, pink, and white. The general topic is flowers. The subtopic is types of flowers. And the colors are specific terms that describe flowers. A topic makes more sense and is more easily understood if you understand how it is broken down into individual ideas and how these ideas are related to one another and to the entire topic.

It is often helpful to organize information visually so that you can see how it all fits together. One technique for describing related ideas is called a **concept map**. In a concept map, an idea is represented by a word or phrase enclosed in a box. There are several ideas in any concept map. A connection between two ideas is made with a line. A word or two that describes the connection is written on or near the line. The general topic is located at the top of the map. That topic is then broken down into subtopics, or more specific ideas, by branching lines. The most specific topics are located at the bottom of the map.

To construct a concept map, first identify the important ideas or key terms in the chapter or section. Do not try to include too much information. Use your judgment as to what is really important. Write the general topic at the top of your map. Let's use an example to help illustrate this process. Suppose you decide that the key terms in a section you are reading are School, Living Things, Language Arts, Subtraction, Grammar, Mathematics, Experiments, Papers, Science, Addition, Novels. The general topic is School. Write and enclose this word in a box at the top of your map.

SCHOOL

Now choose the subtopics—Language Arts, Science, Mathematics. Figure out how they are related to the topic. Add these words to your map. Continue this procedure until you have included all the important ideas and terms. Then use lines to make the appropriate connections between ideas and terms. Don't forget to write a word or two on or near the connecting line to describe the nature of the connection.

Do not be concerned if you have to redraw your map (perhaps several times!) before you show all the important connections clearly. If, for example, you write papers for Science as well as for Language Arts, you may want to place these two subjects next to each other so that the lines do not overlap.

One more thing you should know about concept mapping: Concepts can be correctly mapped in many different ways. In fact, it is unlikely that any two people will draw identical concept maps for a complex topic. Thus there is no one correct concept map for any topic! Even though your concept map may not match those of your classmates, it will be correct as long as it shows the most important concepts and the clear relationships among them. Your concept map will also be correct if it has meaning to you and if it helps you understand the material you are reading. A concept map should be so clear that if some of the terms are erased, the missing terms could easily be filled in by following the logic of the concept map.

SCHOOL
provides courses in
Language Arts   Mathematics   Science
read   learn   includes   teaches about
Novels   Grammar   Living things   perform
write
Papers   Addition   Subtraction   Experiments

SOUND AND LIGHT

The description of amber waves of grain in the familiar song "America the Beautiful" produces a stunning image of golden fields of wheat rhythmically blowing in the breeze. Simply from the use of the word wave, you form a complete picture of the motion of the wheat. It blows back and forth in a smooth, repeating pattern—perhaps not unlike the motion of a flag flying on a windy day. In fact, the expression "Long may it wave" is applied to the flag. You probably use the word wave to describe a variety of motions. You wave hello, wave goodbye, wave a banner, make waves in water.

Beautiful flags of different countries blow rhythmically in the wind.

A variety of appealing sounds is produced as the band marches proudly down the street.

But do you say you hear waves of sound or see waves of light? Probably not. Yet wave motion is necessary to describe both sound and light. Waves are also necessary to describe forms of energy that surround you every day, as well as many technologies that affect your life and your future. In this textbook you will learn about the actual nature of wave motion. You will discover that each type of wave has certain characteristics that make it different from other types of waves, along with characteristics that account for the qualities that make a wave in a wheat field different from a wave of light. And you will become familiar with the wave energy that flies a flag, disturbs a pool of water, warms you and your food, carries sound and light, and transmits information. As you read this textbook, the word wave will become an even more important part of your vocabulary.

Majestic waves rock the ocean waters over and over again without ever stopping.

Discovery Activity

Making Waves

1. Fill a pan or tub with water.

2. Float a narrow instrument such as a pencil on the water.

3. Push the pencil up and down and observe the waves that are made in the water. Draw a wave on a piece of paper.

   ■ What characteristics do you notice about the wave?

   ■ What happens to the waves when they hit the sides of the container?

   ■ What happens if you push the pencil at different speeds?

Characteristics of Waves

Way out in the ocean, far from the eyes of eager surfers, the wind stirs a small ripple into the water's calm surface. As the wind continues to blow, gaining speed and strength, the ripple grows into a full, surging wave. By the time it travels thousands of kilometers to the Hawaiian shore, the wave rises several meters above the surface—forming the famous Hawaiian Pipeline.

Uninterested in its origin and development, the surfers see the wave as the challenge of the day. They run into the water, surfboards in hand. A few quick strokes and they catch the monstrous wave. Most cannot keep ahead of the crushing weight of water for long and are soon pulled under. But one surfer holds on. She steers back and forth as the wave towers over her. The water thunders around her, but success is hers. With a sense of accomplishment and an appreciation of nature's power, she rides the wave all the way to shore.

What is a wave? What do ocean waves have to do with wind? How can a wave travel several thousand kilometers? As you read this chapter, you will find the answers.

Journal *Activity*

You and Your World If you have ever been under an ocean wave as it broke, you know how powerful waves can be. In your journal, describe an experience you have had with any type of water waves—ocean waves, waves in a lake, waves that you made in a pan of water or bathtub. Explain what you were doing and how the waves affected you. Then describe three scenes that illustrate different aspects of ocean waves. Center each description around one of the following words: beauty, destruction, fun.

◄ *Although the powerful wave can't last forever, its awesome energy has given this surfer a wonderful ride.*

Figure 1–1 *A pebble tossed into a still pond creates a disturbance that moves outward along the surface of the water as a wave. The continuous blowing of the wind causes wheat to move in a wavelike pattern.*

1–1 Nature of Waves

Have you ever dropped a pebble into a still pond and observed the circular waves moving outward? Maybe you have watched the waves moving across a field of wheat on a windy day. Or perhaps you have observed huge waves in the ocean during a storm. All these examples illustrate waves. You might be surprised to learn that even light and sound are examples of waves.

Waves and Energy

Think again about ocean waves. Ocean waves continuously roll into the shore, one after the other, night and day. Have you ever wondered how ocean waves can do this without flooding the beaches? The reason is that ocean waves do not actually carry water. As a wave rushes to the shore, the ocean water is moved up and down—but not forward. So even though it looks as if the water itself is moving toward shore, it is actually not. Only the wave moves forward.

To help you understand this, consider another example of waves. What happens to a nearby canoe or pile of leaves in a lake when a motorboat speeds by? The motorboat creates waves that move past the object, causing the object to bob up and down. The waves continue to move forward, but the object remains in approximately the same place. Why? The motorboat disturbs the flat surface of the lake. The disturbance moves outward along the surface of the water as a series of **waves.**

The meaning of the word disturbance should not be new to you. Suppose that you are taking a nap one day in a comfortable hammock when a friend comes along and tilts you out. Your friend has moved you from your resting position. Among other things, you might say that your friend has disturbed you from your rest, or that your friend is a disturbance! After your friend leaves, you return to your nap. In much the same way, particles of water are disturbed from their resting positions by water waves. Once the disturbance has passed, the water particles return to their resting positions. They are not carried by the wave.

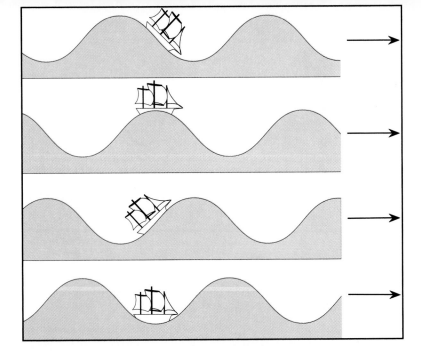

Figure 1–2 *The boat in the water simply bobs up and down as waves roll by. What is a wave?*

What, then, makes up the disturbance? You already learned that it is not matter. The fact that water waves do not carry matter—such as water, canoes, or piles of leaves—is true of any wave. So what then is carried by a wave? Waves carry energy. (Energy is the ability to do work or cause change.) **A wave is a traveling disturbance that carries energy from one place to another.**

Where Do Waves Get Energy?

To understand how waves are made, try making a wave on your own. Tie a rope to a fixed object such as a doorknob or post. See Figure 1–4 on page 14. Jerk the free end of the rope and observe the "bump," or wave motion, that travels to the other end. Now try moving your hand up and down or back and forth over and over again. You have just created a series of waves. How did you do this?

You made the wave motion on the rope by moving your hand up and down or back and forth. Any movement that follows the same path repeatedly is called a **vibration.** You created a vibration. Vibrations are probably quite familiar to you already. The top of a drum vibrates after it is struck by a drumstick. It moves up and down several times, creating

Figure 1–3 *Just the slightest touch, carried like a signal, knocks over a long row of dominoes. Like a falling row of dominoes, a wave can move over a long distance. The substance through which it moves, however, has limited movement.*

Activity Bank

Seeing Sound, p.138

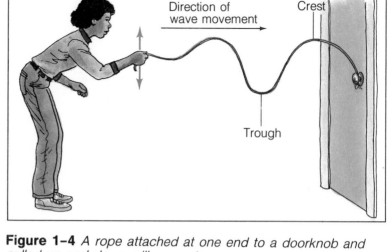

Figure 1–4 *A rope attached at one end to a doorknob and pulled up and down will generate a wave.*

Figure 1–5 *When a piano key is struck, a string is hit and set into vibration. Notice the vibrating string in the center of the photo. Energy is given to the string by the player's finger. What gives energy to the sound wave?*

its characteristic sound. In fact, the vibration of your eardrum in response to sound enables you to hear. A stretched rubber band that is plucked also vibrates for a few seconds. So does a guitar string. Even the Earth can vibrate during a powerful earthquake.

An object that is vibrating is moving. And an object that is moving has energy. A vibrating object gives off some of its energy to nearby particles, causing them to vibrate as well. These particles, in turn, give off energy to the particles next to them, and so on. This movement of energy from a vibrating source outward is a wave. Waves—whether they be ocean waves, waves on a rope, sound waves, or microwaves—have a vibration as their source. The motorboat you just read about gave energy to the water particles in the lake. You gave some of your energy to the rope by moving your hand. Wind blowing back and forth over the ocean creates ocean waves. Electric charges (like the ones you see and feel when you receive a shock) can vibrate to create light and microwaves. Energy is given to the air by the vibration of a guitar string to create a sound wave.

Waves Through Matter and Space

In most of the previous examples, energy from a vibration traveled through a substance. The matter, or substance, through which a wave is transmitted is called a **medium**. Water is a medium for ocean waves. Air is a medium for sound waves. All phases of matter (solid, liquid, and gas) can act as a medium. Waves that require a medium are called **mechanical waves**.

For certain waves, a medium is not required. These waves can be transmitted through a vacuum (space free of particles). Instead of matter, these waves disturb electric and magnetic fields. For this reason, they are called **electromagnetic waves**. Because they do not depend on particles of matter, electromagnetic waves can exist with or without a medium. Light is an electromagnetic wave. Light from the sun can travel to the Earth through the vacuum of space. Light can also travel through air across your room. Microwaves in an oven are electromagnetic waves, as are X-rays used in medicine.

Figure 1–6 *The telescopes of the Very Large Array in New Mexico collect invisible waves that travel with or without a medium. What are these types of waves called?*

1–1 Section Review

1. How are waves and energy related?
2. Where do waves get their energy?
3. What is a medium? A mechanical wave?
4. Describe an electromagnetic wave. How does it differ from a mechanical wave?

Connection—*You and Your World*
5. The programs you watch on television are made up of all sorts of sounds and colors. How do waves make television possible?

1–2 Characteristics of Waves

You just learned that there are many different kinds of waves. Sound waves, light waves, X-rays, microwaves, and ocean waves are but a few examples. All waves, however, share certain basic characteristics. **All waves have amplitude, wavelength, and frequency.**

In order to understand these characteristics of waves, it may help you to represent a wave as a drawing on a graph. The X-axis (the horizontal line)

Guide for Reading

Focus on this question as you read.

▶ *What are the basic characteristics of a wave?*

ACTIVITY

Waves on a Rope

1. Tie one end of a 3-m rope or coiled telephone cord to a stationary object. The knob of a door will work well.

2. Quickly move the other end of the rope up and down. Observe the resulting wave.

3. Increase the speed of the up-and-down movement. What happens to the wave?

4. Increase the height to which you move your arm. What happens to the wave?

■ What is the relationship between the source of a wave and the characteristics of the wave?

represents the normal, or resting, position of the medium, or field, before it is disturbed by a wave. For example, the X-axis might represent a calm sea or a tight rope. The vibrational movements of the wave are shown on the Y-axis (the vertical line). See Figure 1–7. The highest points on the graph are called peaks or **crests.** The lowest points are called **troughs** (TRAWFS).

Amplitude

If the wave disturbs a medium, the particles of the medium are moved from their normal (resting) positions. The distance the particles are moved from their resting positions is shown by the up and down pattern of the graph. Similarly, if the wave disturbs electric or magnetic fields, the graph shows the rise and fall of the fields. In any wave, the amount of movement from rest is shown by the distance above or below the X-axis. The maximum (or greatest) movement from rest is called the **amplitude** (AM-pluh-tood) of the wave. The amplitude can be found by measuring the distance from rest to a crest or from rest to a trough.

The amplitude of a wave indicates the amount of energy carried by the wave. As energy increases, particles of the medium are moved a greater distance from rest. Thus the amplitude of the wave also increases. The amplitude decreases as the wave loses energy. What happens to the crests and troughs of a wave as amplitude increases? As it decreases?

Figure 1–7 *The basic characteristics of a wave are shown here. What is the high point of a wave called? The low point? What does wave amplitude measure?*

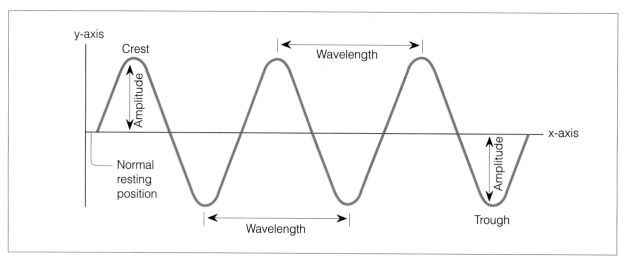

Wavelength

The distance between two consecutive (one after another) crests or troughs of a wave is called the **wavelength**. Actually, the wavelength can be measured from any point on a wave as long as it is measured to the same point on the next wave. Wavelength is usually measured in meters or centimeters. The symbol for wavelength is the Greek letter lambda (λ).

Frequency

The number of complete waves, or complete cycles, per unit of time is called the **frequency** (FREE-kwuhn-see). Because every complete wave has one crest and one trough, you can think of the frequency as the number of crests or troughs produced per unit time. The unit used to measure wave frequency is called the hertz (Hz). This unit is named after Heinrich Hertz, who was one of the first scientists to study certain types of waves. A frequency of one hertz is equal to one wave, or cycle, per second: 1 Hz = 1 wave/sec.

The frequency of a wave depends on the frequency at which its source is vibrating. Think about

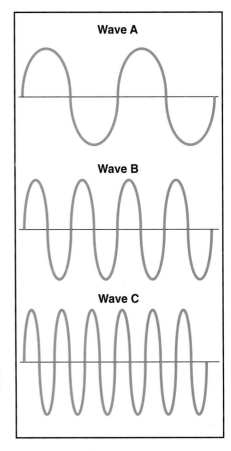

Figure 1–8 *Frequency is the number of complete waves per unit time. What is the frequency of Wave A? Wave B? Wave C?*

Figure 1–9 *Electromagnetic waves of different frequencies produce rainbows. In fact, the same kinds of waves at lower frequencies are used to heat food in microwave ovens. A curious pooch can detect sounds of higher frequencies than you can.*

the rope again. If you move your hand slowly, the rope vibrates slowly. Perhaps you will create one new wave every two or three seconds. If you move your hand rapidly, the rope vibrates rapidly. This way, you may create several waves each second. Try it and see!

Frequency, which is often used to describe waves, is an important characteristic. Frequency is used to distinguish one color of light from another, as well as one sound from another. For example, red light is different from blue light because red light has a lower frequency. A dog can hear a whistle that you cannot hear because dogs can hear sounds at higher frequencies than humans can.

1–2 Section Review

1. Define three basic characteristics of a wave.
2. What is the unit of wave frequency? How is it defined?
3. If the horizontal distance from a crest to a trough is 1.0 m, what is the wavelength?

Critical Thinking—*Making Calculations*
4. Suppose you notice that 20 waves pass a point in 5 sec. What is the frequency? How many waves would pass a point in 1 sec if the wave frequency were two times greater?

Guide for Reading

Focus on this question as you read.

▶ *What is the difference between a transverse wave and a longitudinal wave?*

1–3 Types of Waves

You learned that mechanical waves require a medium through which to travel. Although they share this characteristic, all mechanical waves are not the same. Ocean waves are a different type of wave from sound waves. Why? Although they both transfer energy through a medium, the movement of the disturbance, or wave, through the medium is quite different. Depending on the motion of the medium as compared to the movement of the wave, waves are classified as either transverse or longitudinal.

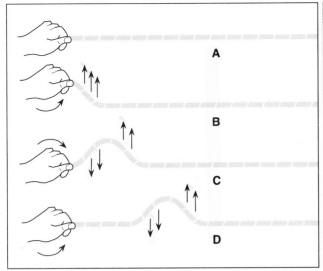

Figure 1–10 *A wave on a rope is the same type of wave as a wave that carries sunlight to your eyes. What type of wave is it?*

Transverse Waves

When one end of a rope is jerked, energy is given to the nearby particles of rope. These particles start to move up and down (vertically) as a result of the energy. As they move, they transfer energy to neighboring particles, which in turn move up and down. As each neighboring rope particle begins to move up and down, energy is transferred from one place to another (horizontally). Each particle moves up and down, but the wave moves horizontally along the rope. Thus the movement of the particles is vertical while the movement of the wave is horizontal. The two movements are at right angles to each other. **A wave in which the motion of the medium is at right angles to the direction of the wave is called a transverse wave.** A wave on a rope is a **transverse wave**. Light and other electromagnetic waves are transverse waves.

Longitudinal Waves

Clap your hands together near your face. Do you hear a clap? Do you also feel air striking your face? When you clap your hands, you move the particles of air away from their resting positions and crowd them together. A space in the medium in which the particles are crowded together is called a compression (kuhm-PREHSH-uhn). Because you give the particles energy, they begin to vibrate back and forth.

ACTIVITY

READING

Mysteries of the Sea

The wonders of the sea have been the topic of many beautiful works of art and literature throughout time. Obtain and read "Sea Songs" by Myra Cohn Livingston. This is a wonderful poem about the magic and mysteries of nature.

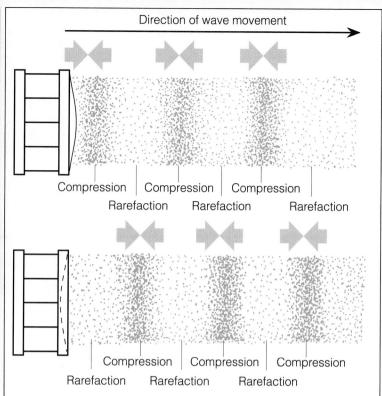

Direction of wave movement

Compression | Compression | Compression

Rarefaction | Rarefaction | Rarefaction

Compression | Compression | Compression

Rarefaction | Rarefaction | Rarefaction

Figure 1–11 *The vibration of a drum head produces compressions and rarefactions in the air. So do the melodious voices of singers in a choir. What type of a wave is a sound wave?*

Figure 1–12 *Longitudinal waves can be represented on a graph. The crests of a longitudinal wave represent the compressions. The troughs represent the rarefactions.*

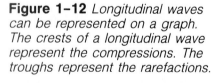

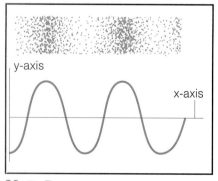

y-axis

x-axis

See Figure 1–11. As the particles of air move to the right, they pass their resting positions and collide with the particles of air next to them. These particles also become compressed. Then the first set of air particles moves to the left while the second set of particles begins to vibrate and moves to the right. This leaves a space that contains many fewer particles. A space in the medium in which there are fewer particles is called a rarefaction (rair-uh-FAK-shuhn).

Each layer of particles pushes the next layer as the compressions move forward through the medium. Each compression is followed by a rarefaction. So rarefactions also move forward. As the layers of particles move back and forth through a medium, compressions and rarefactions develop and move in a regular, repeating way. Energy is transmitted as a wave. A wave that consists of a series of compressions and rarefactions is a **longitudinal** (lahn-juh-TOOD-uh-uhl) **wave.**

As you can see, longitudinal waves are quite different from transverse waves. **In a longitudinal wave, the motion of the medium is parallel to the direction of the wave.** In other words, the particles of the medium move in the same direction in which the wave moves. Sound waves are longitudinal waves.

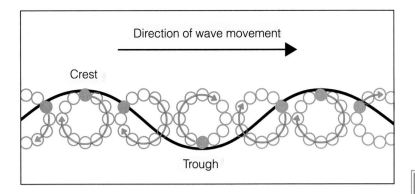

Crest

Direction of wave movement

Trough

Figure 1–13 *The particles affected by a surface wave move in circular patterns. The energy of the wave is transmitted without the movement of the medium as a whole.*

Longitudinal waves can be represented on a graph in the same way transverse waves can. The crests represent the crowded areas, or compressions. The troughs represent the least crowded areas, or rarefactions. The wavelength of a longitudinal wave is the distance between compressions or rarefactions. Frequency is the number of compressions or rarefactions that pass a point per second.

Combinations of Waves

Some waves cannot be described as only transverse or longitudinal. That is because these waves are a combination of the two types of waves. An example of such a wave is a **surface wave**. Surface waves (as their name implies) occur at the surface between two different mediums. Water waves on the surface of the ocean are an example of surface waves. They travel between water and air. The motion of each particle is neither up and down nor back and forth. It is a combination of both movements. The combination produces a wave in which each particle moves in a circle.

1–3 Section Review

1. What is the difference between a transverse wave and a longitudinal wave?
2. What is a compression? A rarefaction?
3. What is an example of a transverse wave? Longitudinal wave?
4. Describe a surface wave.

Critical Thinking—*Making Connections*
5. Describe an example of a transverse wave that gives rise to a longitudinal wave.

1–4 Speed of Waves

Have you ever noticed that you hear thunder several seconds after you see lightning? Even though thunder and lightning are produced at the same time, you see the lightning first because light travels at a much faster speed than sound. Different waves travel at different speeds.

The speed of a wave is determined by the number of waves passing a point in a certain amount of time (frequency) and the length of the wave (wavelength). The speed of a wave is equal to the frequency times the wavelength.

Speed = Frequency × Wavelength

When the frequency of a wave is measured in hertz and the wavelength is measured in meters, the speed of the wave is measured in meters per second.

A wave with a frequency of 4 hertz (Hz) and a wavelength of 2 meters has a speed of 8 meters per second (4 Hz × 2 m = 8 m/sec). If the frequency of the wave were increased to 8 hertz, the wavelength would decrease to 1 meter in the same medium. Why? **In a given medium, the speed of a wave is constant.** Thus the speed must still be 8 meters per second. If the frequency is now 8 hertz, the wavelength must be 1 meter (8 Hz × 1 m = 8 m/sec). An increase in frequency requires a corresponding decrease in wavelength. What would happen to the frequency if the wavelength were increased?

The speed of a wave does not depend on the source or the speed of the source. For example, the speed of sound does not depend on whether it is produced by an airplane or by the snap of your fingers. Nor does the speed of sound depend on the

Figure 1–14 *You see lightning, even at a great distance, before you hear thunder because light travels at a greater speed than sound. Nothing is known to travel faster than the speed of light.*

speed of the airplane that caused it. **The speed of a wave depends upon the medium through which it is traveling.**

One property of a medium that affects the speed of mechanical waves is the density. The density of a medium is a measurement of the medium's mass divided by its volume. A substance that is very dense has more particles or more massive particles than a less dense substance. It may help you to consider some examples: Molasses is more dense than water; water is more dense than air.

If you stir molasses, you will discover that your spoon moves more slowly than it does when you stir water because molasses is more dense than water. The movement of the spoon is slower in the denser medium. The same is true of waves. A wave moves more slowly in a denser medium. The more dense the medium, the slower the speed of a wave in that medium. Why? A denser medium has more inertia to overcome. It is harder to get all of the particles of a denser medium to respond to the energy of the wave and start moving.

Another property of a medium that affects the speed of waves is elasticity. Elasticity refers to the ability of a medium to return quickly to its original shape after being disturbed. A wave moves faster in a more elastic medium. This is because the particles return to their rest positions more quickly.

1–4 Section Review

1. What is the relationship between wave speed, frequency, and wavelength?
2. If the frequency and wavelength of a wave are changed, what happens to the speed? Why?
3. A wave has a frequency of 10 Hz and a wavelength of 30 m. What is its speed?
4. If the frequency of the wave in question 3 were 20 Hz, what would be the wavelength?

Critical Thinking—*Applying Concepts*
5. At 25°C, the speed of sound in air is 346 m/sec. At 0°C, the speed of sound in air is 332 m/sec. Explain why the speed decreases as the temperature decreases.

ACTIVITY
DOING

Waves of Straw

1. Stretch out 2m of masking tape, sticky side up. At right angles to the tape, place a soda straw every 6 cm. When all the straws are in place, place another piece of tape over the straws on top on the first piece of tape.

2. Attach one end of the tape to a stationary object. a doorknob or ring stand will do.

3. Rotate the free end by turning your wrist. Observe the motion of the straws.

What is the role of the straws and tape?

4. Now move the first straw up and down. With a stopwatch, measure the time it takes for the wave to travel to the end of the tape. (Speed = Distance/Time)

5. Attach paper clips to the ends of the straws. Repeat step 4.

How does the speed of this wave differ from the speed of the wave in step 4. Why?

1–5 Interactions of Waves

You learned that waves traveling in the same medium move at a constant speed and in a constant direction. What would happen, however, if the waves encountered a different medium, reached an obstacle, or met another wave? Depending on the conditions, the waves would interact in a certain way. **The four basic wave interactions are reflection, refraction, diffraction, and interference.**

Reflection

Have you ever seen water waves strike a rock or the side of a swimming pool? What happens? You are correct if you have observed that the waves bounce back. When a wave strikes a barrier or comes to the end of the medium it is traveling in, at least part of the wave bounces back. Figure 1–15 shows waves striking a barrier and bouncing back. This interaction is called **reflection** (rih-FLEHK-shuhn). Reflection is the bouncing back of a wave after it strikes a boundary that does not absorb all the wave's energy.

In order to better describe and illustrate reflection (as well as other wave interactions), certain diagrams are used. A line drawn in the direction of motion of a wave is called a ray. Rays are used to show wave activity. The incoming wave is called

Figure 1–15 *According to the law of reflection, the angle of incidence equals the angle of reflection. The green laser beam dramatically illustrates this point as it bounces off three mirrors before it enters the jar.*

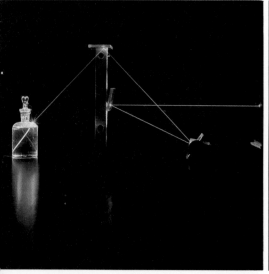

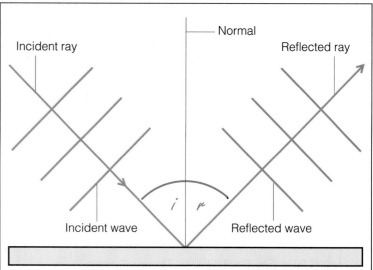

Normal

Incident ray

Reflected ray

Incident wave

Reflected wave

i *r*

Figure 1–16 *Thanks to the reflection of sunlight from the surface of the moon, you can enjoy a moonlit night.*

an incident wave. The wave that is bounced back is called a reflected wave. The angle formed by the incident ray and an imaginary line drawn perpendicular to the barrier is called the angle of incidence (i). The angle formed by the same perpendicular line and the reflected ray is called the angle of reflection (r). The line drawn perpendicular to the barrier is called the normal. **The law of reflection states that the angle of incidence (i) is equal to the angle of reflection (r).** This means that the angle the incident wave makes with a surface is equal to the angle the reflected wave makes with the same surface.

You are probably already familiar with reflection in your everyday life. For example, when you look in the mirror, you are taking advantage of the reflection of light to see yourself. Concert halls and theaters are designed to use reflection to make the sound more powerful. An echo is another example of reflection.

Refraction

Waves do not bend as they travel through a medium unless an obstacle gets in the way. Waves travel in straight lines. The light from your flashlight, for example, travels straight across the room. However, when waves pass at an angle from one medium to another—air to water or glass to air, for example—they bend. The waves bend because the speed of the waves changes as the waves travel from one medium to another. Remember, wave speed is constant only for a particular medium. As a wave enters a different medium, its speed changes.

The bending of waves due to a change in speed is called **refraction** (rih-FRAK-shuhn). Refraction occurs because waves move at different speeds in different mediums. As waves pass at an angle from one medium to another, they may speed up or slow down.

You can easily see for yourself the results of the refraction of light by trying the following activity. Place a pencil diagonally in a glass of water. What do

Sound Around, p.139

Handball Physics

1. Choose a spot on a wall. Roll a rubber ball directly to that spot so that the direction of movement of the ball is perpendicular to the wall. How is the ball reflected?

2. Change your position and again roll the ball to the spot, but this time at an angle to the wall.

3. Change your position again and roll the ball to the wall from a different angle.

■ Do your observations confirm the law of reflection?

■ How can you improve your handball or racquetball game?

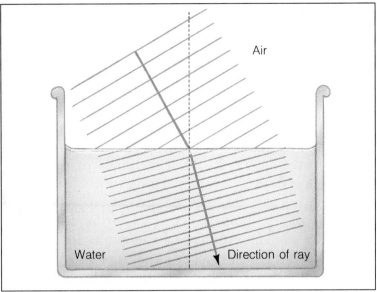

Figure 1–17 *Waves bend as they pass from air into water. Notice how the light beam in the photograph demonstrates refraction.*

Figure 1–18 *The refraction of light as it passes from one medium to another makes this flower stem look as if it were broken. Why does refraction occur?*

you see? The pencil appears to be split into two pieces. The light waves traveling through air are slowed down and bent when they travel through water. Here's another way to observe refraction. Place a coin in the bottom of an empty cup. Move the cup so that the coin is out of your line of sight. Then fill the cup with water. Does the coin become visible? Explain what has happened.

Diffraction

If you duck around a corner to escape an oncoming snowball, you avoid getting hit. The snowball will continue past you. But if you try the same tactic to avoid someone who is yelling at you, you can still hear the shouting. Why? Sound travels as a wave, whereas the snowball does not. And although waves travel in straight lines through a medium, if an obstacle is encountered, waves bend around it somewhat and pass into the region behind it.

The bending of waves around the edge of an obstacle is called **diffraction** (dih-FRAK-shuhn). Diffraction is a result of a new series of waves being formed when the original waves strike an obstacle. The amount of diffraction depends on the wavelength and the size of the obstacle.

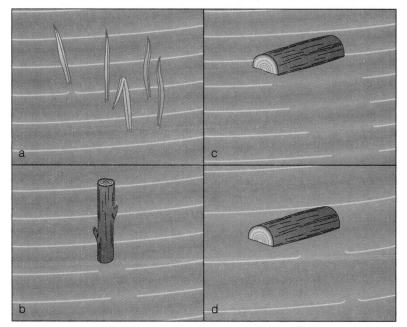

Figure 1–19 *The larger the wavelength of the waves compared with the size of the object, the more the waves will bend around the object, or diffract. In frames a, b, and c the wavelength of the waves is the same, but less bending occurs as the size of the object increases. Frames c and d show the same object, but because the wavelength is greater in frame d, more bending is observed.*

Presto Chango, It's Gone, p.140

Interference

Suppose that you and a friend are holding the ends of a piece of rope. You both snap the ends at the same time. Two waves are sent toward each other. What will happen when the waves meet at the middle of the rope?

When two or more waves arrive at the same place at the same time, they interact in a process called **interference.** The waves combine to produce a single wave. To see how two waves combine, the displacement, or distance from rest, of one wave is added to the displacement of the other wave at every point along the wave. If the displacement of one wave is below the resting position (the x-axis), it is subtracted from the displacement of the other wave. Waves can combine in two different ways.

CONSTRUCTIVE INTERFERENCE If waves combine in such a way that the disturbance that results is greater than either wave alone, constructive interference occurs. For example, if the crests of one wave meet the crests of the other wave, constructive interference occurs. The crests of the two waves add together to form a single wave. The amplitude of the single wave is equal to the sum of the amplitudes of the two original waves.

Rough Riding

Assemble the following materials: 2 rubber or wooden wheels, firmly attached to an axle, and a large piece of velvet cloth or course sandpaper.

1. Roll the wheels and axle across a smooth tabletop. Describe the direction of motion.

2. Place the velvet cloth or course sandpaper on the tabletop.

3. Roll the wheels again but in such a way that only one wheel moves across the rough surface. Describe the motion.

4. What effect does this motion have on the direction of the wheels? What causes this to happen?

■ What does this activity illustrate about wave interactions?

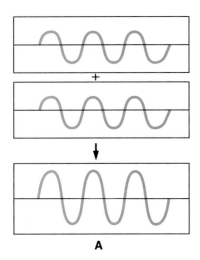

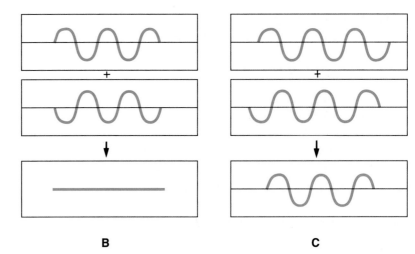

A   B   C

Figure 1–20 *When two waves arrive at the same place at the same time, they interfere with each other. If the same points on each wave combine, they add together (A). If opposite points combine, they cancel out (B). And if the waves meet at different points, some parts add and some parts subtract, causing the wave to vary (C).*

DESTRUCTIVE INTERFERENCE If waves combine in such a way that the disturbance that results is less than either wave alone, destructive interference occurs. For example, if the crests of one wave meet the troughs of the other wave, destructive interference occurs. The crests and troughs combine by subtracting from each other to form a single wave. The amplitude of this wave is the difference between the amplitudes of the original waves.

If the crest of one wave occurs at the trough of the other wave and both waves have the same amplitude, the waves will cancel each other out. The combination of the waves will result in no wave at all.

Standing Waves

If you continuously shake the end of a rope up and down, the waves you create will travel to the opposite end of the rope and be reflected back. After a few vibrations, there will be quite a jumble of waves. But if you vibrate the rope at just the right frequency (move your hand up and down a certain number of times each second), something interesting happens. A wave that does not appear to be moving results. This type of wave is called a **standing wave**.

The points on a standing wave where destructive interference results in no energy displacement are called nodes (NOHDZ). Nodes are located on the X (resting) axis. The points at which constructive interference causes maximum energy displacement are called antinodes. Any crest or trough may be called an antinode.

Standing waves can occur at more than one frequency for waves on a rope. The lowest frequency that produces a standing wave causes the first pattern shown in Figure 1–21. Twice that frequency will cause two loops. Three times that frequency will cause three loops, and so on for any multiple of the frequency.

The frequency at which a standing wave occurs is called the natural frequency, or **resonant frequency**, of the object. When an object is vibrating at its natural frequency, little effort is required to achieve a large amplitude. Perhaps you have pushed a friend or a child on a swing. At first it may be difficult to push the swing. But once the swing reaches its natural frequency of moving back and forth, you need only push lightly to keep the swing moving.

An object that is vibrating at its natural frequency can cause a nearby object to start vibrating if that object has the same natural frequency. The ability of an object to vibrate by absorbing energy of its own natural frequency is called resonance (REZ-uh-nuhns). A singer can shatter glass by singing a clear, strong high-pitched note. If the natural frequency of the glass is the same as the natural frequency of the note sung by the singer, the glass will vibrate due to resonance. Glass is not very flexible. So if it absorbs enough energy, it will shatter.

You are already familiar with other examples of resonance—although you may not realize it! You are applying the principle of resonance every time you tune in your radio or turn the station on your television. Each station broadcasts at a specific frequency.

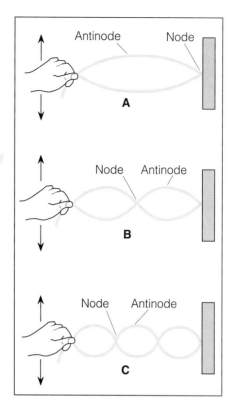

Figure 1–21 *Standing waves occur when an object vibrates at its resonant frequency and multiples of that frequency. How many nodes are there in the standing wave that occurs at four times the resonant frequency of this rope?*

Figure 1–22 *Raindrops on a lake show what happens when waves interfere. At the regions where the circular waves overlap, the waves exhibit greater amplitudes than in the center of each circle.*

Figure 1-23 *When the tuning fork on the left is set in motion, it begins to vibrate at its natural frequency. These vibrations travel through the air and the wooden resonance box, which strengthens them. The tuning fork on the right will begin to vibrate "in sympathy" because it absorbs energy at its own natural frequency. Resonance caused the collapse of the Tacoma Narrows Bridge in November 1940.*

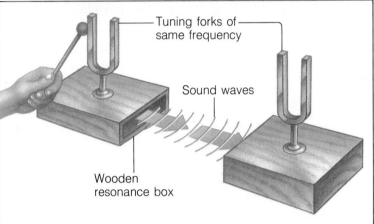

Tuning forks of same frequency

Sound waves

Wooden resonance box

When you tune into a station, you are matching the frequency of your radio or television with the frequency of the broadcasting station.

1–5 Section Review

1. Name the four wave interactions.
2. What is reflection? What is the law of reflection?
3. What is refraction? Diffraction?
4. Compare constructive and destructive interference.
5. What is a standing wave?

Critical Thinking—*Applying Concepts*

6. If you drop a perfectly transparent piece of glass into perfectly clear water, you can still see the glass. Why?

Shake, Rattle, and Roll

About one million times a year, some part of the Earth's surface shakes and trembles in a violent release of energy. Such an event is called an *earthquake*. An earthquake occurs when energy stored in the Earth's crust builds up to the extent that it causes different sections of rock to move. From the point at which the earthquake originates, waves ripple out in all directions. Earthquake waves are known as seismic waves.

There are three different types of *seismic waves*. Those that travel the fastest are called primary waves, or P waves, because they are recorded first. P waves are longitudinal waves. They move through solids, liquids, and gases. The next type of seismic wave is a secondary wave, or S wave. S waves are recorded after P waves. S waves are transverse waves that move through solids but not through liquids and gases. The last type of seismic wave is a surface wave, or L wave. When P waves and S waves reach the surface, they are converted into L waves. L waves travel along the Earth's surface, causing the ground to move up and down. L waves cause the most damage because they are responsible for bending and twisting the Earth's surface.

Scientists have learned a great deal about earthquakes by studying seismic waves. By using an instrument called a seismograph, which records seismic waves, scientists can determine the presence and strength of an earthquake. The greater the amplitude of the seismic waves, the greater the amount of energy they carry—and thus the more powerful the earthquake.

Studying earthquake wave activity has enabled scientists to gain insight into predicting and describing earthquakes and also to think about preventing them someday.

Notice the destruction caused by seismic waves that traveled across the surface of the Earth during the 1989 California earthquake.

Laboratory Investigation

Observing Wave Properties of a Slinky®

Problem

What are the characteristics of a wave?

Materials   *(per group)*

Slinky® or other coiled spring

Procedure

1. On a smooth floor, stretch the spring to about 3 m. Have one person hold the spring at each end. **CAUTION:** *Do not overstretch the spring.*
2. Make a loop at one end of the spring, as shown in the accompanying figure.

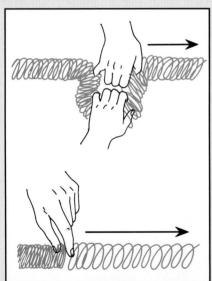

3. Release the loop and observe the motion of the wave. Observe the reflected wave.
4. Move one end of the spring back and forth on the floor. Draw a diagram of the wave you observe.
5. Repeat step 4, increasing the rate at which you move the spring back and forth.
6. Now squeeze together the first 20 cm of the spring, as shown in the accompanying figure.
7. Release the compressed section of the spring and observe the wave as it moves down the spring.

Observations

1. In step 3, is the reflected wave on the same or the opposite side as the original wave?
2. What happens to the frequency when you increase the rate at which the spring is moved back and forth?
3. What happens to the wavelength when you increase the rate at which the spring is moved back and forth?

Analysis and Conclusions

1. Are the waves generated in steps 1 through 5 transverse or longitudinal? Explain your answer.
2. Are the waves generated in steps 6 and 7 transverse or longitudinal? Explain your answer.
3. What is the relationship between the rate at which the spring is moved back and forth and the frequency? And the wavelength?
4. What are three characteristics of a wave?
5. **On Your Own**   How are the properties of ocean waves related to the wind that causes them? Design an investigation to test your answer.

Study Guide

Summarizing Key Concepts

1–1 Nature of Waves

▲ A wave is a disturbance that carries energy from one place to another.

▲ Mechanical waves require a medium. Electromagnetic waves can travel through a vacuum or through a medium.

1–2 Characteristics of Waves

▲ When a wave is graphed, the highest points are called crests and the lowest points are called troughs.

▲ The maximum displacement of a wave is its amplitude.

▲ The distance between two consecutive similar points on a wave is its wavelength.

▲ The frequency of a wave is the number of waves that pass a point per unit time.

1–3 Types of Waves

▲ In a transverse wave, the motion of the medium is at right angles to the direction of the motion of the wave.

▲ The motion of the medium is parallel to the direction of a longitudinal wave.

1–4 Speed of Waves

▲ Wave speed equals frequency times wavelength.

▲ The density and elasticity of a medium affect the speed of a wave.

1–5 Interactions of Waves

▲ Reflection occurs when a wave strikes a barrier and bounces back.

▲ The law of reflection states that the angle of incidence equals the angle of reflection.

▲ Refraction is the bending of waves due to a change in speed.

▲ The bending of waves around the edge of a barrier is called diffraction.

▲ Waves traveling through the same space at the same time interfere with each other.

▲ A wave produced at the resonant frequency of a material is a standing wave.

Reviewing Key Terms

Define each term in a complete sentence.

1–1 Nature of Waves
wave
vibration
medium
mechanical wave
electromagnetic wave

1–2 Characteristics of Waves
crest
trough
amplitude
wavelength
frequency

1–3 Types of Waves
transverse wave
longitudinal wave
surface wave

1–5 Interactions of Waves
reflection
refraction
diffraction
interference
standing wave
resonant frequency

Chapter Review

Content Review

Multiple Choice

Choose the letter of the answer that best completes each statement.

1. A wave transports
   - a. energy.
   - b. matter.
   - c. water.
   - d. air.

2. An example of an electromagnetic wave is a
   - a. water wave.
   - b. light wave.
   - c. wave on a rope.
   - d. sound wave.

3. The maximum displacement of a wave is measured by its
   - a. wavelength.
   - b. frequency.
   - c. amplitude.
   - d. speed.

4. The distance between two consecutive troughs is the
   - a. frequency.
   - b. amplitude.
   - c. medium.
   - d. wavelength.

5. Light travels as a
   - a. transverse wave.
   - b. mechanical wave.
   - c. longitudinal wave.
   - d. wave in a medium.

6. In a given medium, if the frequency increases,
   - a. wavelength increases.
   - b. speed increases.
   - c. speed stays the same.
   - d. speed decreases.

7. The bending of waves due to a change in speed is called
   - a. diffraction.
   - b. refraction.
   - c. reflection.
   - d. interference.

8. The bending of waves around the edge of an obstacle is called
   - a. diffraction.
   - b. reflection.
   - c. refraction.
   - d. interference.

9. The interaction of waves that meet at the same point is called
   - a. reflection.
   - b. diffraction.
   - c. interference.
   - d. refraction.

True or False

If the statement is true, write "true." If it is false, change the underlined word or words to make the statement true.

1. Waves transfer matter.
2. A vibration gives rise to a wave.
3. The energy carried by a wave is indicated by its wavelength.
4. In a transverse wave, the particles of the medium move parallel to the direction of the wave.
5. Frequency is measured in hertz.
6. When a wave strikes an obstacle, part of the wave is reflected.
7. The incoming wave is the incident wave.
8. When two waves combine to subtract from each other, constructive interference occurs.

Concept Mapping

Complete the following concept map for Section 1–1. Refer to pages R6–R7 to construct a concept map for the entire chapter.

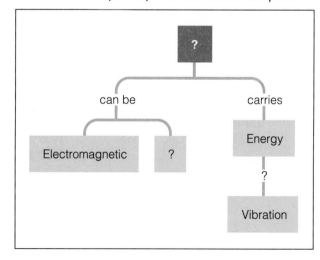

Concept Mastery

Discuss each of the following in a brief paragraph.

1. Explain how waves are related to energy and vibrations.
2. Why can microwaves travel both through empty space and through air and food?
3. A pendulum makes 32 vibrations in 20 sec. What is its frequency?
4. Suppose a red spot is painted at one point on a spring. Describe what happens to the red spot as a transverse wave passes by. Describe what happens as a longitudinal wave passes by.
5. Must a wave be transverse or longitudinal? Explain.
6. How does the density and the elasticity of a medium affect the speed of a wave?
7. Describe what happens to an incident wave after it hits an obstacle at an angle.
8. Explain the wave activity that enables you to hear a marching band before it rounds the corner.
9. Two waves occur in the same place at the same time, yet no wave can be detected there. Explain how this can be true.

Critical Thinking and Problem Solving

Use the skills you have developed in this chapter to answer each of the following.

1. **Applying concepts** Explain how a long line of people at a concert can be a medium for a longitudinal wave.
2. **Making inferences** Satellites above the Earth that are used to relay communications such as radio and television must rely on electromagnetic waves. Why?
3. **Identifying relationships** Red light has a longer wavelength than blue light. Why must it have a lower frequency?
4. **Identifying relationships** Energy is neither created nor destroyed. How does this law require that waves be reflected when their medium ends abruptly?
5. **Making calculations** Complete the following table.
6. **Applying concepts** Short races in a track meet are run over straight courses. Timers at the finish line start their watches when they see smoke from the starting gun, not when they hear the shot. Why? (The speed of sound in air is 340 m/sec.)
7. **Applying concepts** If you watch people trying to carry a large pan full of water, you will see that some are quite successful at it while for others, who are equally as careful, the water sloshes around badly. What makes the difference?
8. **Using the writing process** You are a sportswriter with a background in science. While covering an exciting game one Sunday, you cannot help but notice the enthusiasm of the crowd. Quite often throughout the game, entire sections of people stand up and sit down again in groups, from left to right, in an activity called "The Wave." Write a short newspaper article on the crowd's participation.

Speed (m/sec)	Frequency (Hz)	Wavelength (m)
150	2.0	
	250	1.5
200		0.5
	200	1.0

Sound and *Its Uses*

Guide for Reading

After you read the following sections, you will be able to

2–1 What Is Sound?
- Describe sound and its transmission.

2–2 Properties of Sound
- Identify the properties of sound.

2–3 Interactions of Sound Waves
- Classify wave interactions of sound.

2–4 Quality and Sound
- Distinguish between noise and music.

2–5 Applications of Sound
- Describe some uses of sound.

2–6 How You Hear
- Describe how sound is heard.

It is another hot, dry day on the African plain. All is quiet. Among the many animals that make their home in this region is a large herd of elephants. Silently the elephants walk along single file—all, that is, except the young ones who run alongside to keep up with their mothers. The herd's destination is uncertain—perhaps it is a watering hole or an area with a fresh supply of food. Their motion is slow but constant. Then suddenly, for no obvious reason, all the elephants freeze in place. Some stop with their trunks up in the air. Others halt in midstep. But just as quickly and unexpectedly as they stopped, they resume motion. Only this time, they make a sharp turn to head in a different direction.

You may be wondering why you are reading about elephants in a chapter on sound. After all, what did their behavior have to do with sound? It was done in silence. Or was it? Actually, elephants can communicate over several kilometers using sounds that cannot be heard by humans. As you read this chapter you will learn about sound—what it is, how it is made, and why you cannot hear the language of the elephants.

Journal *Activity*

You and Your World Go to a quiet place. Close your eyes for a minute and just listen. You will probably find out that it is not as quiet as you think. You may hear a light fixture humming above your head, a truck passing in the distance, or a clock ticking on the wall. You are almost always surrounded by sounds. In your journal, describe the sounds you hear.

◀ *African elephants, of assorted sizes, out for an afternoon stroll*

2–1 What Is Sound?

Have you ever noticed that if you lean on a piano while it is being played, you can feel the beat of the music? Or if you place your hand over a loud-speaker, you can actually feel a sensation in the air directly in front of it? And did you also know that if you could take your radio to the moon, you would not be able to hear it (or any noise, for that matter)? In order to understand these facts, you must learn about the nature of sound.

How Sounds Are Made

A bell shaking rapidly, a drum moving up and down, and a harp string bouncing back and forth are all examples of objects that make sounds. What do these examples have in common? They each vibrate when they are making sounds. As an object vibrates, it gives energy to the particles of matter around it. The energy causes the particles of matter to vibrate as well—and in such a way that a series of compressions (crowded areas) and rarefactions (less-crowded areas) moves outward from the source. Recall from Chapter 1 that a moving series of compressions and rarefactions is called a longitudinal wave. **Sound, which is produced when matter vibrates, travels as a longitudinal wave.**

Anything that vibrates produces sound. Your vocal cords, for example, vibrate to produce sound. When you speak, air from your lungs rushes past

Figure 2–1 *Although a tuning fork does not appear to vibrate, you can clearly see that it does from the splashes it makes when placed in water. If there are no particles of a medium to vibrate, there will be no sound. Why can't the astronaut riding along in a lunar vehicle hear the engine?*

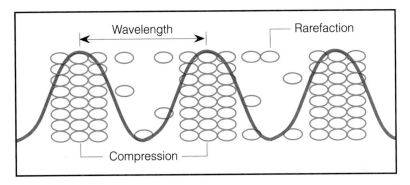

Figure 2-2 *A longitudinal wave is composed of a series of compressions and rarefactions moving through a medium. The compressions form the crests of the wave. The rarefactions form the troughs. What is the distance between consecutive crests called?*

your vocal cords (two folds of tissue located in your throat), causing them to vibrate. As your vocal cords move inward, the air between them is pushed together, forming a compression. As your vocal cords move outward, an area with fewer particles of air is left, creating a rarefaction. A series of compressions and rarefactions travels outward from your vocal cords, making up the sound of your voice. When you speak, therefore, it is not the air you breathe that travels to the receiver; it is the sound waves that travel. The particles of air simply move back and forth.

Think about your radio again. The mechanism in the speaker of your radio actually moves the air in front of it by causing the particles to vibrate. If there were no air (medium) in front of the speaker, there could be no sound. Without a medium to transmit vibrations, there can be no sound. On the surface of the moon, where there is no atmosphere, there is no medium to transmit sound. Thus there can be no sound on the moon or in any vacuum.

Perhaps you are familiar with the age-old question, "If a tree falls in the forest and no one is present to hear it, is there a sound?" Although some people might answer no, you would probably say yes now that you know about sound. A falling tree sends a disturbance through the air and the ground. This disturbance travels as a longitudinal wave—a sound wave. Thus a sound is produced.

Speed of Sound

You learned in Chapter 1 that the speed of a wave depends on the properties of the medium and not on the source of the wave. For example, if a flute and a trumpet are both played in the same orchestra, both sounds will travel through the same

Activity Bank

Bells A' Ringing, p.141

Viewing Vibrations

Sounds are caused by vibrations. You can observe this by experimenting with a tuning fork.

1. Strike the prongs of a tuning fork with a pencil and then hold the fork close to your ear. What happens? What happens when you touch the prongs of the fork?

2. Again strike the prongs of a tuning fork and place the ends of the prongs in a glass of water. What happens?

3. Tie a small piece of cork to a string and hold the string in one hand so the cork can swing freely. Strike the prongs of a tuning fork and hold one prong against the cork. Observe what happens.

Figure 2–3 *The pioneering flight of Chuck Yeager made possible the now-commonplace flights of airplanes at more than three times the speed of sound.*

Figure 2–4 *Vibrations of the vocal cords cause air particles to vibrate and thus produce sound. This is how the sounds you make when you speak, sing, or yell are produced. It is also the way the operatic artist Jessye Norman produces the beautiful notes of the aria. Norman is an accomplished singer with the New York Metropolitan Opera.*

air and arrive at your ear at the same time. Both sounds will travel at the same speed. **The speed of sound is determined by the temperature, elasticity, and density of the medium through which the sound travels.**

TEMPERATURE One important characteristic of the medium that determines the speed of sound is temperature. Lowering the temperature of a substance makes the motion of the particles more sluggish. The particles are more difficult to move and slow to return to their original positions. Thus sound travels slower at lower temperatures and faster at higher temperatures.

In 1947, Captain Chuck Yeager took advantage of the relationship between the speed of sound and temperature to set a historic record. Captain Yeager was the first person to fly faster than the speed of sound. When he "broke the sound barrier," he was flying at a speed of 293 meters per second. But if the speed of sound in air averages about 340 m/sec (faster than he was traveling), how could Yeager have broken the sound barrier? Yeager was flying at an altitude of 12,000 meters. At this altitude, the temperature is so low that the speed of sound is only 290 m/sec—3 m/sec/sec less than the speed achieved by Yeager. A vehicle on the ground would have to travel about 50 m/sec faster to beat the speed of sound.

ELASTICITY AND DENSITY Although most sounds reach you by traveling through air, sound waves can travel through any medium. To return to the example of the piano, sound waves cause the particles in the wood of the piano to vibrate. That is what you feel when you lean on the piano. Years ago, Native Americans put their ears to the ground in order to find out if herds of buffalo or other animals were nearby. By listening for sounds in the ground, they could hear the herds much sooner than if they listened for sounds in the air. The speed of sound in the ground at 20°C is 1490 m/sec—more than four times as fast as in air. Can you think of an example where you listen through some material other than air?

Perhaps from these examples you have realized that sound travels at its greatest speed in solids and at its slowest speed in gases. What is it about the phase of the medium (solid, liquid, or gas) that determines the speed of sound? For one thing, the speed depends on the elastic properties of the medium. This means that if the particles of the substance are disturbed, they must be able to return to their original positions easily.

To understand this, consider the following: You decide to run 1 kilometer down a paved road. You are able to run quickly and steadily because your feet spring off the solid blacktop. By the time you get to the end, you are hardly tired. This is similar to sound traveling through materials in which the particles are rigidly bound together and return to place quickly. Sound travels fastest in elastic materials such as these. But suppose you run the same distance on a wet beach where each time you step, the sand sinks under your feet. This time you have to put more energy into each step. When you finish this jog, you will be more tired. This is what happens to sound waves in less-elastic mediums. The sound waves travel more slowly and lose energy more quickly.

Solids are generally more elastic than either liquids or gases. The particles in a solid do not move very far and bounce back and forth very quickly as the compressions and rarefactions of a sound wave go by. Thus sound travels more easily through solids than it does through liquids and gases. Most liquids are not too elastic. Sound is not transmitted as well in liquids as it is in solids. Gases are even more inelastic than liquids. So gases are the poorest transmitters of sound.

In materials in the same phase of matter, the speed of sound is slower in the denser material.

SPEED OF SOUND	
Substance	Speed (m/sec)
Rubber	60
Air at 0°C	331
Air at 25°C	346
Cork	500
Lead	1210
Water at 25°C	1498
Sea water at 25°C	1531
Silver	2680
Copper	3100
Brick	3650
Wood (Oak)	3850
Glass	4540
Nickel	4900
Aluminum	5000
Iron	5103
Steel	5200
Stone	5971

Figure 2–5 *The speed of sound varies in different mediums. In what medium does sound travel the fastest? The slowest?*

Activity Bank

Just Hanging Around, p.143

Figure 2–6 *A piano can be made to send melodious tunes through the air to your ear. But the sound waves also travel through the wood of the piano and can be felt as vibrations. Does sound travel faster through wood or through air? Why?*

Because the denser medium has greater mass in a given volume, it has more inertia. Its particles do not move as quickly as those of the less-dense material. The speed of sound in dense metals such as lead and gold is much less than the speed of sound in steel or aluminum. Lead and gold are also less elastic—another reason why the speed of sound is slower in these metals.

2–1 Section Review

1. What is sound? What kind of wave carries sound?
2. What characteristics of the medium affect the speed of sound?
3. Compare the transmission of sound in solids, liquids, and gases.

Connection—*Earth Science*

4. Light travels faster than sound. Thunder and lightning occur at the same time. However, thunder is heard after a flash of lightning is seen. How can the time in between the two be used to calculate how far away a storm is?

Guide for Reading

Focus on this question as you read.

▶ *How are the characteristics of sound related to the physical characteristics of the wave?*

2–2 Properties of Sound

Now you know that all sounds originate in the same basic way. They are produced by vibrations and transmitted as longitudinal waves. Yet there are millions of different sounds in everyday life—each having certain characteristics that make it unique. Think about the many sounds you hear every day. How you hear and describe a sound depends on the physical characteristics of the sound wave. As you have read in Chapter 1, the physical characteristics of a wave are amplitude, frequency, and wavelength. These are the factors that determine the sounds you hear.

Frequency and Pitch

Certain sounds are described as high, such as those produced by a piccolo, or low, such as those produced by a bass drum. A description of a sound

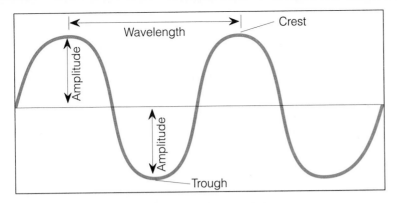

Figure 2–7 *Some of the basic characteristics of sound can be seen in this wave diagram. What three characteristics can you identify?*

as high or low is known as the **pitch** of the sound. The pitch of a sound depends on how fast the particles of a medium vibrate. Each complete vibration—one compression and one rarefaction—makes up a wave. So the pitch of a sound depends on the number of waves produced in a given time. This is a definition of frequency. **Thus the pitch of a sound depends on the frequency of waves.**

Sound waves that have a high frequency are heard as sounds of high pitch. A violin produces high-pitched sounds. Sound waves that have a low frequency are heard as sounds of low pitch. A tuba produces low-pitched sounds. A high note sung by a soprano may have a frequency of 1000 hertz. You will recall that a frequency of 1 hertz is equal to one wave, or cycle, per second. Thunder has a low pitch—its frequency is less than 50 hertz.

Frequency is an especially important characteristic of sound because the ear can respond to only certain frequencies. The normal human ear is capable of detecting from about 20 to 20,000 vibrations per second, or hertz. Sounds with frequencies higher than 20,000 hertz are called **ultrasonic** (uhl-truh-SAHN-ihk) because they are above the range of human hearing. Some animals hear quite well at this level. If you have ever used a dog whistle, you know that when you blow on it, your dog comes running even though you do not hear any sound. Dogs can hear sounds with frequencies up to 25,000 hertz. Or perhaps you have seen a cat appear startled although you have heard nothing. Cats can hear sounds with frequencies up to 65,000 hertz. So the cat probably did hear a noise! Porpoises can hear sounds with frequencies up to 150,000 hertz. And bats actually produce ultrasonic sounds and then use the echoes

Figure 2–8 *The pitch of a sound depends on the frequency of the waves. How would you describe the frequency of the high-pitched sound of a flute? Of the low-pitched sound of a sousaphone?*

Figure 2–9 *Your pet may be aware of activities you do not notice because it can hear sounds that you cannot. Similarly, dolphins can hear and communicate with one another by using sounds with frequencies more than seven times those heard by humans.*

Figure 2–10 *Certain natural phenomena such as volcanoes and earthquakes emit sounds with frequencies lower than humans can hear. Many animals become aware of the danger of such Earth movements before humans do because the animals hear the infrasonic warnings.*

to locate prey or to avoid bumping into objects. In this way, bats can fly about safely in the dark.

Sound waves with frequencies below 20 hertz are referred to as **infrasonic** (ihn-fruh-SAHN-ihk) because they are below the range of human hearing. You might have wondered earlier why, if sound is produced by all vibrations, you do not hear a sound from vibrations such as your arm swinging back and forth. The truth is, you actually do produce a sound, but the frequency is much too low for your ear to hear it. Think back to the elephants you read about in the beginning of this chapter. Elephants communicate by infrasonic sounds. Earthquakes and volcanoes emit infrasonic sound waves. Even certain kinds of machinery vibrate at frequencies in the infrasonic region. This can be dangerous to workers exposed to these vibrations for long periods of time, because although the sound is not heard, the energy carried by the waves can alter body processes over time.

Doppler Effect

There is another common occurrence that depends on the frequency of sound waves. You experience it quite often without even realizing it. Have you ever stood on the side of the road as a police car sped by with its siren on? If so, then you

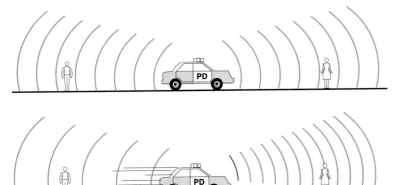

Figure 2–11 *When there is no movement between the source of a sound and its receiver, the sound waves are sent out equally in all directions. When there is motion, however, the sound waves either pile up or spread out, depending on the motion. This changes the frequency of the sound. How does this affect the pitch of the sound?*

should have noticed that the pitch of the siren was higher as it approached you and then lower as it moved away from you. Or perhaps you have been in a moving car that passed a fire station as the alarm was sounding. You may recall a similar change in the pitch of the alarm as you drove toward and then away from the station. This change in pitch is referred to as the **Doppler effect**. The Doppler effect occurs whenever there is motion between the source of a sound and its receiver. Either the source of a sound or the receiver must move relative to the other.

Consider again the police car moving toward you. Each time the siren sends out a new wave, the car moves ahead in the same direction as the wave. This causes the waves to be pushed together. Because they are pushed together, the waves have shorter wavelengths and higher frequencies as they reach you, the receiver. Higher frequency results in higher pitch. As the car passes you, it is traveling in the opposite direction of the sound waves that reach you. Thus the waves are more spread out, causing lower frequency and a lower pitch to the sound. Can you think of some other examples in which you experience the Doppler effect?

Intensity and Loudness

What is the difference between the sound of thunder and the sound of a handclap? You might say that one sound is louder than the other. Do you know what makes one sound louder than another?

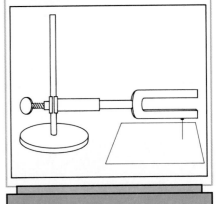

Figure 2–12 *Decibel levels of some familiar sounds are shown in this chart. Which sounds would be considered painful?*

INTENSITY OF SOUND

Sound	Decibels	Sound	Decibels
Threshold of human hearing	0	Heavy street traffic	70–80
Rustling leaves	10	Vacuum cleaner	75–85
Whisper	10–20	Loud music	90–100
Very soft music	30	Rock concert	115–120
Classroom	35	Threshold of pain	120
Average home	40–50	Jet engine	170
Conversation	60–70	Rocket engine	200

Perhaps this example will help. If you pound your fist lightly on a table top, you hear a sound. But if you strike the table top much harder, you hear a louder sound. By striking the table top harder, you put more energy into the sound. Loudness is related to the amount of energy carried by a wave.

The amount of energy carried by a wave in a certain amount of time is called the **intensity** of the wave. As you may recall from Chapter 1, the energy of a wave is indicated by the amplitude. The larger the amplitude, the greater the intensity. For sound waves, this means that the compressions are more crowded and the rarefactions are less crowded. **Intensity determines the loudness of a sound.** The greater the intensity of a sound, the louder the sound is to the ear. Thunder sounds louder than a handclap because the intensity of thunder is much greater than the intensity of a handclap.

A scale has been developed to measure the relative intensity of sounds. The scale is based on a unit called the decibel (dB). A sound with an intensity level of 0 decibels is so soft that it can barely be heard. Thunder, on the other hand, has an intensity level of 120 decibels. Thunder is a tremendously loud sound. Sounds with intensities greater than 120 decibels can actually cause pain in humans.

A jet engine has an intensity level of about 170 decibels. This sound is really painful to human ears. So ground crews exposed to this sound must wear special ear protection to avoid injuring their ears.

Figure 2–13 *Elevated trains and subways produce sounds with high intensity levels. What are some other high-intensity sounds that you encounter every day?*

Music with an intensity above 85 decibels also can cause ear damage. Many rock stars wear ear plugs at their own concerts to prevent hearing loss.

2–2 Section Review

1. What characteristic of sound is determined by frequency? By amplitude?
2. Compare ultrasonic and infrasonic waves.
3. Describe the Doppler effect.

Critical Thinking—*Applying Concepts*
4. How might the Doppler effect be used to measure velocity?

2–3 Interactions of Sound Waves

Have you ever found yourself tapping your foot to the beat of a good tune? Perhaps you have even said that you liked one song better than another because it had a faster or a slower beat. Beats result from interactions between waves. As you learned in Chapter 1, waves that occur at the same place at the same time combine, or interfere. Because sound is a wave, sound experiences interference.

Combining Sounds

Interference of sound waves, like the interference of any other waves, can be constructive or destructive. **When sound waves combine in such a way that the resulting disturbance is greater than either wave alone, constructive interference occurs.** As a result of constructive interference, the intensity of a sound is increased. The sound is louder. Outdoor amphitheaters use band shells to amplify (increase the loudness of) the music through constructive interference. The sound waves produced by the orchestra or the singers bounce off the ceiling and walls so that they join together.

Guide for Reading

Focus on this question as you read.

▶ *What happens when sound waves combine?*

Figure 2–14 *Sound waves from these loudspeakers interfere constructively at point A but destructively at point B. Why is this important when setting up a radio or stereo system?*

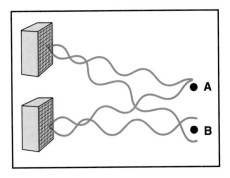

What Was That?

Airplanes that travel faster than the speed of sound produce loud, unpleasant noises called sonic booms. Read "Sonic Boom" by John Updike to find out what one famous writer thinks about this type of sound.

If an amphitheater is not designed properly, the resulting sound waves heard by parts of the audience will have smaller amplitudes than those produced. **When sound waves combine in such a way that the resulting disturbance is less than either wave alone, destructive interference occurs.** As a result of destructive interference, the intensity of a sound is decreased. The sound is softer.

Destructive interference can produce spots in which sounds cannot be heard at all. These areas are referred to as dead spots. Dead spots are especially troublesome in large halls that have hard surfaces which reflect (bounce back) sounds into the room. The reflected sounds interfere destructively and cancel each other. Engineers who work in **acoustics,** or the science of sound, try to design concert halls and auditoriums with no dead spots. Acoustical engineers must carefully design the shape, position, and materials of an auditorium to eliminate interference problems and prevent the sound waves from being absorbed.

You may still be wondering where beats come from. If two sources of sound are close in frequency but not exactly the same, the sound waves they produce combine in such a way that the amplitude and loudness of the resulting wave changes at regular intervals. At some points the interference is constructive and at others it is destructive. It gets louder and softer at intervals that depend upon the difference in frequencies. The repeated changes in loudness are called beats.

In addition to giving music its characteristic tune, beats have useful applications as well. For example, a

Figure 2–15 *Beats occur when two waves of slightly different frequencies add together. What happens at the beat?*

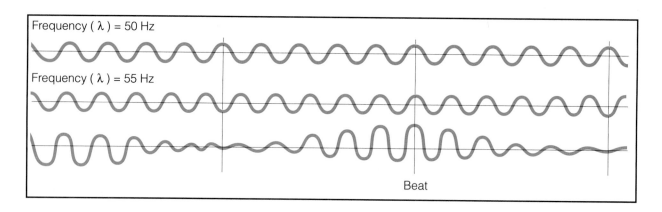

Frequency (λ) = 50 Hz

Frequency (λ) = 55 Hz

Beat

Figure 2–16 *Even the Early Greeks knew about the importance of acoustics when they designed this theater in Italy. Although material and construction have improved, as illustrated by the Sydney Opera House in Australia, architects and acoustical engineers must still design theaters to take advantage of the action of sound waves.*

piano tuner strikes a tuning fork and the matching string on a piano at the same time. If beats exist between the two sound waves, the piano tuner knows that the string is vibrating at a different frequency from the tuning fork. When the string is adjusted so that the beats disappear, the string is tuned to the same frequency as the tuning fork.

2–3 Section Review

1. What happens when two sound waves combine if the crest of one wave meets the crest of the other? If the trough of one wave meets the crest of the other?
2. What is the difference between constructive and destructive interference in terms of sound intensity?
3. What are beats? How are they produced?

Connection—*You and Your World*
4. Why does putting a carpet and furniture in a room decrease the amount of sound in the room?

2–4 Quality and Sound

A jackhammer vibrates to produce sound, yet you would probably not call it music. Most likely, you would call it noise! You might call the squeak of chalk across a chalkboard noise too (and it probably is not one of your favorite sounds). A talented musician, however, makes an instrument produce melodious, pleasing music. What is the difference between music and noise? The difference is in the quality of the sound.

Sound Quality

Every different sound has its own quality. You would not mistake a flute for a trumpet even when they are both playing the same note. This is because the instruments have a different **sound quality**. Sound quality is also called **timbre** (TAM-ber).

Sounds are produced by a source vibrating at a certain frequency. In reality, however, most objects that produce sounds vibrate at several different frequencies at the same time. Each frequency produces a sound with a different pitch. The blending of pitches gives the sound its timbre.

In Chapter 1 you learned that every object has a natural, or resonant, frequency. When the object vibrates at that frequency, or multiples of that frequency, it produces a standing wave. The lowest frequency at which this occurs is the fundamental. The note that is produced is called the **fundamental tone**. The fundamental tone has the lowest frequency and pitch possible for that object. While the whole object (a string, for example) is vibrating at its resonant frequency, sections of the string are vibrating faster than the fundamental tone and producing sounds with higher frequency and pitch. The sounds of higher frequencies are called **overtones**. Sounds always have a fundamental tone and one or more overtones. **The blending of the fundamental tone and the overtones produces the characteristic quality, or timbre, of a particular sound.**

Without overtones, a trumpet and a flute would sound exactly the same. In fact, a violin, a clarinet, and your friend's voice would all have the same

Figure 2–17 *A fundamental tone is produced when the whole string vibrates. When sections of the string vibrate faster than the fundamental tone, notes called overtones are produced. How does the pitch of the overtones compare with that of the fundamental tone?*

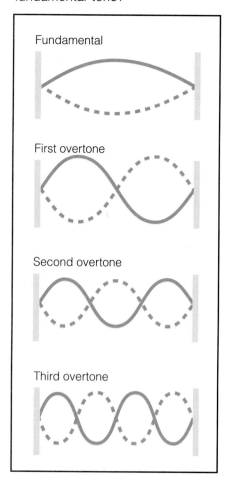

Fundamental

First overtone

Second overtone

Third overtone

sound quality if it were not for overtones. Sound quality is so unique for each person's voice that voiceprints have been used to identify a person.

Music

Now that you know about sound quality, you can understand the difference between music and noise. If the relationship between the fundamental tone and the overtones results in a sound with a pleasing quality and a clear pitch, the sound is considered music. **A sound is music if it has a pleasing quality, a definite identifiable pitch, and a repeated timing called rhythm.**

The design of a musical instrument is responsible for establishing the relationship between the fundamental and the overtones. In musical instruments, as in any source of sound, a vibration must exist. The material that vibrates can vary. Percussion instruments (drums, bells, and cymbals) are made to vibrate by being struck. Stringed instruments (guitars and violins) are either plucked or rubbed to produce regular vibrations. In woodwind and brass instruments (flutes, trombones, pipe organs), columns of air are made to vibrate.

Consider a stringed instrument such as a guitar. When a string is plucked, it begins to vibrate. The vibration produced has a certain frequency and thus a certain pitch. But by changing the length, tightness, or thickness of the string, the string can be made to produce different pitches.

A shorter string vibrates at a higher frequency and produces a sound of higher pitch. The short strings of a ukulele (yoo-kuh-LAY-lee) produce higher pitched notes than the longer strings of a cello. The vibrating length of a string can be changed by the proper positioning of the fingers along the string. Musicians do this to produce the pitches they desire.

The tighter the string, the higher the frequency of vibration and the higher the pitch. A stringed instrument is tuned either by tightening or loosening each string on the instrument. If a string is tightened, it produces a higher pitched sound. If it is loosened, it produces a lower pitched sound.

Strings of different masses also have different sounds. Because the speed of a wave on a heavy

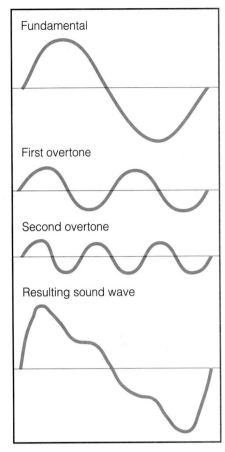

Fundamental

First overtone

Second overtone

Resulting sound wave

Figure 2–18 *At any point, the amplitude of the fundamental wave is added to those of the overtones. The result is the actual sound wave.*

Making Music

1. Fill six empty bottles, each with a different amount of water.

2. Blow across the top of each bottle and observe the sound that is produced. Compare the pitches.

How could you change the pitch of the bottles without changing the amount of water in the bottles?

Figure 2–19 *Musical instruments are made in a variety of ways. A pipe organ and unusual horns native to Peru are made to produce music when a column of air vibrates. The Korean instrument known as a kayakum makes music by the plucking of strings. And when the Guatemalan musician strikes his marimba made of gourds, unique sounds are produced.*

Figure 2–20 *This construction worker is producing sound, but it certainly is not music. It is noise. How is noise different from music?*

string is less than on a lighter string, a sound on a thicker string will have a lower frequency for the same wavelength. So thicker strings produce lower pitched sounds. The strings on a bass guitar are thicker than the strings on a lead guitar. The strings on a piano or a harp are of different lengths and masses. What kinds of sounds do thinner strings produce?

In woodwind and brass instruments, columns of air vibrate. Like waves on a string, the waves in air produce a sound with a pleasing quality. The characteristics of the sound are determined by the medium—the length of the air column and its temperature. Have you ever blown air across the top of a bottle to make a noise? When you do this, you set up waves in the air inside the bottle, thereby creating a sound.

Noise

So why is the squeak of chalk on a chalkboard so unpleasant? The answer is quite simple: It is noise.

Noise has no pleasing quality, no identifiable pitch, and no definite relationship between the fundamental tone and the overtones. A sound such as that made by a jackhammer or a piece of chalk on a chalkboard does not have a clear pitch or a repeating pattern. It does have a certain quality, but it is not a pleasing one. Rather than a small collection of frequencies, all of which are multiples of the fundamental, noise has a mixture of frequencies that have no relation to one another.

When noise reaches a level that causes pain or damages body parts, it becomes noise pollution. Noise pollution has become a major concern of society because noise pollution can have a serious effect on people's health. The stress from listening to loud noises over a prolonged period of time can damage delicate tissues of the ear.

What can be done about noise pollution? In many countries, laws have been passed to prohibit noise pollution. People are not allowed to bring loud radios into public places. Of course, you don't have to wait for a law to follow this example. You can look for sources of noise pollution in your home. Then you can determine which sources are within your control. You may not want to stop using a loud appliance in your kitchen. But by placing a rubber pad under the appliance, you can lessen its noise level. In what other ways can you help prevent noise pollution?

DISCOVERING

Changing Pitch

1. Obtain a stringed instrument such as a guitar or a violin. Notice the relative thickness of each string.

2. Pluck each string. Which string has the lowest pitch?

3. Pluck one of the strings again and quickly place your finger on the string at the halfway position. What happens to the pitch when you shorten the string like this?

4. Pluck another string and tighten the tension of the string with the tuning knob. What happens to the pitch when the tension is increased?

■ The pitch of a male voice becomes lower during a period of development known as puberty. What must happen to the vocal cords at this time?

2–4 Section Review

1. Describe sound quality.
2. Explain how music is produced by a stringed instrument. By a woodwind instrument. By a percussion instrument.
3. What is the difference between music and noise?

Critical Thinking—*Classifying Information*
4. Sounds with dB levels between 60 and 100 can be annoying. Sounds above 100 dB can cause damage to hearing. Classify the following sounds as either annoying or damaging: snowmobile, food blender, power mower, jet plane, loud rock band, subway train, police siren.

2–5 Applications of Sound

When you think about the importance of sound in your life, different thoughts probably come to mind. The voices that are part of your daily interactions with people, the hustle and bustle of a busy city, the birds chirping in a country meadow, and the songs you enjoy listening to or singing are just a few examples. But in addition to making life more colorful and enjoyable, sound also has some important applications.

Sonar

Perhaps you think of bats only as scary creatures of the night. But bats display an important use of sound. Bats live and are most active in total darkness, yet they can reach a desired target with pinpoint accuracy. (And they never bump into the walls of the cave!) How do they manage this? Bats send out high-frequency sound (ultrasonic) waves as they fly. The waves bounce off objects such as walls or insects or mice and reflect back to the bat. Such reflected sound is called an echo. Bats use echoes to determine the location of dinner and to navigate around the black interior of their cave.

In this way bats are similar to many ships and submarines. Ships also use sound waves to navigate and to locate objects in the dark depths of the ocean. Suppose, for example, that the cargo of a

Figure 2–21 *You may find them a little unnerving, but bats are actually interesting creatures. They use sound waves to "see." If their eyes are covered, their travel is not impaired at all. But if their ears are covered, they are helpless. What is the use of sound for navigation called?*

ship sinks in the middle of the ocean. The water is far too deep to send divers down in search of the lost merchandise. How then can the cargo be located so that it can be retrieved? **High-frequency ultrasonic waves are used in a system called *Sound Navigation And Ranging*, or sonar.**

A research ship on the ocean surface sends a sound wave into the water. The sound wave travels in a straight line until it reaches a barrier, such as the ocean floor or the sunken cargo. When it hits the barrier, it is reflected back to the ship as an echo. The time it takes for the wave to travel down to the bottom and back up is carefully measured. Using this time and the speed of sound in ocean water, researchers are able to calculate the distance the wave traveled (Distance = Speed × Time). By dividing the total distance by two (half going down and half going up), the depth of the barrier can be determined. If the calculated distance is less than the distance shown on a map of the ocean floor, researchers know that the cargo is located in that spot. (The cargo reflected the sound wave, not the ocean floor.)

Sonar devices are often used in commercial fishing to locate large schools of fish. Sonar is also used to find oil and minerals within the interior of the Earth.

Have you ever seen or used a camera that adjusts the focus automatically? Such a camera may use sonar. Some cameras send ultrasonic waves toward the subject being photographed. After the echo

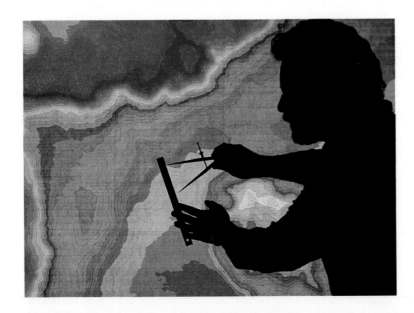

Figure 2–22 *More than a colorful diagram, this map of the ocean made by sonar shows elevation, obstacles, and temperature. Why is a map of the ocean floor useful?*

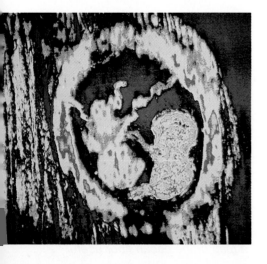

Figure 2–23 These twins don't know it, but they have had their first picture taken—even before they are born. Ultrasonic waves enable doctors to observe the development of unborn babies, as well as the condition of other structures within the body.

returns to the camera, the camera calculates the distance to the subject and sets the proper focus.

A type of sonar is presently being tested in cars. A system in which sonar is used to calculate the distance from a car to nearby objects such as curbs and other vehicles could be used to prevent accidents and to enable cars to park more easily and safely.

Ultrasonic Cleaning

Some objects are too delicate to clean with rough sponges or harsh detergents. Are you surprised to learn that you can clean these objects with sound? Ultrasonic waves are used to clean jewelry, electronic components, and delicate machine parts. To do this, the object is placed in a mild liquid. Sound waves are then sent through the liquid, causing it to vibrate with great intensity. The vibration knocks the dirt off the object without harming it.

Sound and Medicine

A technique much like sonar is used in medicine to diagnose medical problems. An ultrasonic wave is directed into the body. Its reflection from barriers within the body—such as organs and bones—is then detected. The technique is referred to as ultrasound. Using this technique, abnormal growths or pockets of fluid can be discovered.

Images much like X-rays are produced during an ultrasound procedure. But unlike X-rays, which are taken at one instant, ultrasonic waves can be used continuously to show motion occurring within the body. Perhaps you have seen an ultrasonic picture of a developing fetus (unborn baby). Doctors actually watch the movement of the fetus for several minutes at a time on a screen similar to the screen of a television set. Ultrasound has the added advantage that normal ultrasonic waves do not alter body cells.

In addition to allowing doctors to "see" inside the body, ultrasonic waves can be used to treat certain medical conditions. Sometimes unwanted tissue must be destroyed. By focusing ultrasonic waves of high intensity on the unwanted material, larger than normal amounts of energy are deposited on the material, causing it to be destroyed. Ultrasound is also used in physical therapy to provide local heating of injured muscles.

2-5 Section Review

1. What is sonar? What does sonar stand for? How does sonar work?
2. How are ultrasonic waves used in medicine?
3. What are some advantages of ultrasound over X-rays?
4. The speed of sound in ocean water is 1530 m/sec. If it takes 3 seconds for a sound wave to make a round trip from a sonar device, what is the distance to the reflecting object?

Connection—*Earth Science*
5. What are some of the advantages of using sonar to locate oil and mineral deposits in the Earth?

2-6 How You Hear

Guide for Reading

Focus on this question as you read.

▶ *How does the body detect sounds?*

Do you remember the famous question posed at the beginning of this chapter about a tree falling in the forest? You learned that even if no one is around to hear the tree fall, a sound is still produced. Sound is a form of energy that causes particles of a medium to vibrate back and forth.

If the question had been whether a sound is heard, the answer would have been no. In order for a sound to be heard, three things are needed. One, there must be a source that produces the sound. Two, there must be a medium to transmit the sound. And three, there must be an organ of the body that detects the sound. **In humans, the organ of the body that detects sound is the ear.**

How do you hear a series of compressions and rarefactions? Look at Figure 2–24 on page 58. Hearing begins when sound waves enter the **outer ear**. The outer ear acts as a funnel for the waves. The waves move through the ear canal and strike a lightly stretched membrane called the **eardrum**. The vibrating air particles cause the eardrum to vibrate much like a musical drum.

Vibrations from the eardrum enter the **middle ear**. The middle ear contains the three smallest bones in the body. The first bone, the hammer, picks up the vibrations from the eardrum. The hammer

Figure 2-24 *The illustration shows the structure of the human ear. What are the three main parts of the ear?*

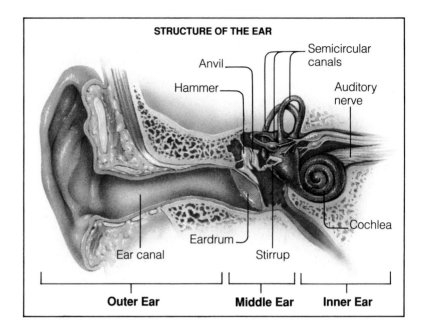

STRUCTURE OF THE EAR

Anvil
Hammer
Semicircular canals
Auditory nerve
Cochlea
Eardrum
Ear canal
Stirrup

Outer Ear   **Middle Ear**   **Inner Ear**

passes the vibrations to the second bone, the anvil. The anvil transmits the vibrations to the third bone, the stirrup. The stirrup then sets another membrane vibrating. This membrane transmits the vibrations to a liquid-filled **inner ear**.

The vibrations in the inner ear are channeled into the **cochlea**. The cochlea, which is shaped like a snail shell, contains a liquid and hundreds of cells attached to nerve fibers. The nerve fibers join together to form one nerve that goes to the brain. The cells detect movements in the liquid of the cochlea and convert them to electrical impulses. The nerve fibers transmit the electrical impulses to the brain, where they are interpreted as sound.

2-6 Section Review

1. Where does hearing begin in humans?
2. What are the three main parts of the ear?
3. What is the function of the eardrum?
4. Where are the nerve impulses interpreted as sound?

Connection—*Life Science*

5. Sometimes an extremely loud sound actually tears the eardrum. Explain what characteristic of the sound wave causes the injury and how the injury affects hearing.

CONNECTIONS

Seeing Sound

"Oooouuut" yells the umpire as he throws his right hand wildly up over his shoulder, his thumb stretched outward. A little dramatic, you might think as you dust yourself off. But hand signals in baseball have some important origins.

At the turn of the twentieth century, a deaf professional baseball player played outfield. Because he was deaf, he could not hear the umpires' calls. To help him, several hand signals, or signs, were developed. These signals enabled him to "see" the calls and follow the game. A great variety of signs are still used in baseball today.

Silent signals are certainly not limited to baseball. In fact, complex languages of manual symbols and gestures are used for communication in many cultures—in particular, among the deaf community. These languages are known as *sign languages.* Sign language combines signs, gestures, facial expressions, and body movements to communicate ideas. For example, by cradling your right arm in your left arm to rock an imaginary infant, you make the sign for baby. By positioning your hand as though tipping the brim of a baseball cap, you create the sign for boy. And by moving your closed hand, thumb extended, along your jaw line as if tying an imaginary bonnet string, you

make the sign for girl. For many proper nouns, names, addresses, and words that have no sign, a system called finger spelling is used. Finger spelling involves using hand positions to represent the letters of the alphabet.

Just as one culture may communicate by speaking English, Spanish, French, or any other language, the deaf community communicates by using sign language. One language of signs, the American Sign Language, is used by so many people that it is the fourth most common language in the United States. But signing is even more than a means of communication. It is actually a beautiful form of expression.

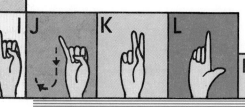

Laboratory Investigation

Investigating Properties of Sound Waves

Problem

What are sympathetic vibration, resonance, interference, beats, and the Doppler effect?

Materials *(per group)*

```
2 tuning forks, 320 Hz
rubber band
resonance box
400-mL beaker filled with water
```

Procedure 🧪

1. *Vibration in a medium:* Strike a tuning fork against the heel of your shoe and then insert the prongs into a beaker of water. Observe what happens.

2. *Sympathetic vibration:* Strike a tuning fork against the heel of your shoe and bring it within a few centimeters of a second tuning fork with the same frequency. Place the second tuning fork a few centimeters from your ear. Observe what happens.

3. *Resonance:* Strike a tuning fork against the heel of your shoe and note the loudness of the sound. Strike the tuning fork once again and then touch the base of its stem to the top of the resonance box. Note the loudness of the sound.

4. *Interference:* Strike a tuning fork against the heel of your shoe and bring one of the prongs 2 or 3 centimeters from your ear. Slowly rotate the tuning fork completely. Carefully note any change in the loudness of the sound.

5. *Beats:* Fasten a rubber band securely on the middle of one prong of a tuning fork. Obtain a second tuning fork of the same frequency. Strike both forks against the heel of your shoe and place the bases of the stems of the forks on the resonance box. If the sound is constant, reposition the rubber band and try again. Carefully note the sound emitted by the forks.

6. *Doppler effect:* Strike a tuning fork against the heel of your shoe with an extra force. Rapidly move the tuning fork in a wide arc round your head. The effect can best be heard by a second observer several meters away. Note what you hear.

Observations

1. In which steps of the procedure did the loudness of the sound change?
2. In which steps of the procedure did the pitch of the sound change?

Analysis and Conclusions

1. What does sound do to its medium?
2. What is sympathetic vibration? How does it happen?
3. Why does sound increase with the use of a resonance box?
4. Is sound louder during constructive or destructive interference?
5. How are beats produced?
6. When the tuning fork was moving in a circle, the pitch alternately increased and decreased. Did it increase when the fork moved toward you or away from you?

Study Guide

Summarizing Key Concepts

2–1 What Is Sound?

▲ Sound is a form of energy that causes particles of a medium to vibrate back and forth.

▲ Sound is a longitudinal wave.

▲ The speed of sound depends on the properties of the medium. Sound travels faster at higher temperatures, in more elastic mediums, and in less dense mediums.

2–2 Properties of Sound

▲ The frequency of a sound wave determines the pitch of the sound.

▲ The Doppler effect is a change in the frequency and pitch of a sound due to the relative motion of the source or the observer.

▲ The amplitude of a sound wave determines its intensity. The loudness of a sound depends on the intensity of the wave.

2–3 Interactions of Sound Waves

▲ Sound waves can interfere constructively, making the sound louder, or destructively, making the sound softer.

▲ Two sources of sound that have slightly different frequencies produce beats.

2–4 Quality and Sound

▲ A sound has quality, or timbre. The blending of the fundamental tone and overtones produces quality.

▲ Music has a clear relationship among the fundamental tone and the overtones. Noise does not.

2–5 Applications of Sound

▲ Sound navigation and ranging, or sonar, is used to measure distances in the ocean, in the Earth, and in the body.

2–6 How You Hear

▲ The ear is the human organ that detects sound.

▲ The outer ear collects sound waves and sends them into the ear canal. The waves cause the eardrum to vibrate.

▲ The vibrations of the eardrum are transmitted to three small bones in the middle ear and then into the fluid in the cochlea.

▲ Special cells in the cochlea change the vibrations into electric nerve impulses that are sent to the brain by nerve fibers.

Reviewing Key Terms

Define each term in a complete sentence.

2–2 Properties of Sound
pitch
ultrasonic
infrasonic
Doppler effect
intensity

2–3 Interactions of Sound Waves
acoustics

2–4 Quality and Sound
sound quality
timbre
fundamental tone
overtone

2–5 Applications of Sound
sonar

2–6 How You Hear
outer ear
eardrum
middle ear
inner ear
cochlea

Chapter Review

Content Review

Multiple Choice

Choose the letter of the answer that best completes each statement.

1. Sound travels as a(an)
   a. transverse wave.
   b. electromagnetic wave.
   c. light wave.
   d. longitudinal wave.
2. The speed of sound is fastest in a
   a. vacuum.   c. liquid.
   b. gas.   d. solid.
3. An increase in the speed of sound may be due to an increase in
   a. temperature.   c. amplitude.
   b. density.   d. pitch.
4. Pitch is related to
   a. frequency.   c. speed.
   b. interference.   d. amplitude.
5. The loudness of a sound depends on its
   a. frequency.   c. Doppler effect.
   b. amplitude.   d. pitch.

6. The quality of sound depends on its
   a. amplitude.   c. speed.
   b. loudness.   d. overtones.
7. The lowest frequency at which an object can vibrate is called its
   a. timbre.   c. fundamental
   b. resonance.   tone.
                       d. overtone.
8. Using sound to measure distance is called
   a. sonar.   c. cochlea.
   b. resonance.   d. acoustics.
9. Vibrations from the eardrum enter the
   a. outer ear.   c. inner ear.
   b. middle ear.   d. brain.
10. The hammer, anvil, and stirrup are in the
    a. outer ear.   c. cochlea.
    b. inner ear.   d. middle ear.

True or False

If the statement is true, write "true." If it is false, change the underlined word or words to make the statement true.

1. The source of a sound is a <u>vibrating</u> object.
2. The speed of sound in hydrogen gas is <u>faster</u> than in oil.
3. An increase in elasticity <u>decreases</u> the speed of sound.
4. As the energy of a sound wave increases, the <u>frequency</u> of the wave also increases.
5. The loudness of a sound depends on the <u>intensity</u> of the sound wave.
6. <u>Constructive</u> interference of sound waves may result in dead spots.
7. <u>Noise</u> has a definite repeating pattern.

Concept Mapping

Complete the following concept map for Section 2-1. Refer to pages R6-R7 to construct a concept map for the entire chapter.

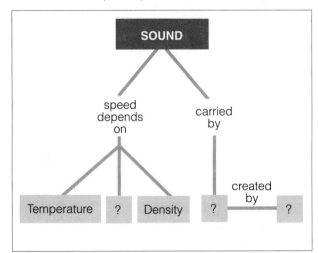

Concept Mastery

Discuss each of the following in a brief paragraph.

1. Describe why a bell ringing inside a vacuum chamber cannot be heard.
2. If you sit in the last row of an auditorium during a concert, why might you see the drummer hit the drum before you actually hear it?
3. Explain why your voice and the voices of your friends do not sound alike.
4. Trace the path of the sound of a hand-clap from the moment the clap is made to the moment you interpret the sound.
5. How come you do not hear a sound when you bat your eyelashes?
6. Suppose that a few pipes of a pipe organ are next to a heating unit. How would this affect the performance of the organ?
7. Describe the sound of a car whose horn is stuck as it approaches you and then passes you.
8. Describe the factors that affect the sound of a guitar string.

Critical Thinking and Problem Solving

Use the skills you have developed in this chapter to answer each of the following.

1. **Making comparisons** List the following materials from best to worst as transmitters of sound: (a) iron, (b) oxygen gas, (c) soup. Explain the reasoning behind your list.
2. **Making graphs** Using the following data, plot a graph showing how the speed of sound in air varies with temperature:

Temperature (in °C)	Speed (in m/sec)
−10	325
0	331
10	337
20	343

3. **Interpreting graphs** From your graph in question 2, determine the speed of sound in air at 18°C and at 25°C. By how much does the speed of sound change if the temperature changes 1°C?
4. **Identifying relationships** Draw a wave diagram to illustrate each of the following sounds: (a) high pitched and loud, (b) low pitched and soft, (c) low pitched and loud.
5. **Relating concepts** Sometimes a whistling sound is heard in a room when a window is slightly open on a windy day. How is this observation related to the principle of a wind instrument?
6. **Designing an experiment** How could two observers on opposite banks of a river use sound to measure the river's width?
7. **Applying definitions** Overtones that sound good together are said to be in harmony. In order for sounds to be harmonic, their overtones must have frequencies that are whole-number multiples of the fundamental. Which of the following frequency combinations will produce harmonic sounds? What will the other combinations produce? (a) 256, 512, 768, 1024 Hz; (b) 128, 256, 1024 Hz; (c) 288, 520, 2048 Hz; (d) 128, 288, 480 Hz; (e) 512, 1024, 4096 Hz
8. **Using the writing process** Use what you have learned about sound and your imagination to write a short story entitled "The Day There Was No Sound."

Light and the Electromagnetic Spectrum

Guide for Reading

After you read the following sections, you will be able to

3-1 Electromagnetic Waves

- Describe the properties of electromagnetic waves.
- Explain how electromagnetic waves are produced.

3-2 The Electromagnetic Spectrum

- Identify the regions of the electromagnetic spectrum.
- Describe the uses of electromagnetic waves of different frequencies.

3-3 Visible Light

- Explain three ways by which luminous objects produce light.

3-4 Wave or Particle?

- Describe the particle nature of electromagnetic waves.

Bzzzz. Bzzzz. A determined honeybee descends on a delicate white flower for a delicious meal of nectar. No big deal, you might think. But in the world of plants, this is an extremely important event.

When a honeybee lands on a flower to drink nectar, pollen sticks to the bee. When the honeybee flies to another flower to feast again, some of the pollen falls on that flower. Thus the honeybee pollinates the plant without even realizing it.

The trick is to get the honeybee to fly to a flower. Nature has devised some clever methods for attracting honeybees to flowers. Brightly colored flower petals are one such method. Why, then, in a field of fabulously colored flowers, does the honeybee choose a white one? The answer is that some flowers have markings on them that are invisible to humans—like a secret code for bees only! These markings are seen only in ultraviolet light. Ultraviolet markings direct the honeybee to the center of the flower—the source of nectar.

What is ultraviolet light, and how is it produced? What other types of rays exist around you? You will find the answers to these and other questions in this chapter.

Journal *Activity*

You and Your World Different types of lights are an important part of your everyday life that you may perhaps take for granted. Imagine what it must have been like to live before electric lights were invented. In your journal, describe how your life might have been different.

A hungry honeybee gathers nectar and pollen from a tiger lily flower.

3–1 Electromagnetic Waves

What does sunlight have in common with the X-rays used in a doctor's office? Are you surprised to learn that they are both waves? They're not matter waves that you can feel or hear. They are electromagnetic waves. You may remember reading about electromagnetic waves in Chapter 1. Although you might not realize it, you are constantly surrounded by thousands of electromagnetic waves every day. Sunlight (visible light) and X-rays are only two types of electromagnetic waves. Other types are radio waves, infrared rays, ultraviolet rays, and gamma rays.

Nature of an Electromagnetic Wave

An electromagnetic wave, as its name suggests, is both electric and magnetic in nature. An electromagnetic wave consists of an electric field and a magnetic field. These fields are not made up of matter like that in a football field or a soccer field. Electric and magnetic fields are the regions through which the push or pull of charged particles and magnets is exerted. (Charged particles and magnets can push or pull certain other objects without even touching them.) **An electromagnetic wave consists of an electric field and a magnetic field positioned at right angles to each other and to the direction of motion of the wave.** See Figure 3–2. Because the electric and magnetic fields are at right angles to the direction of motion of the wave, electromagnetic waves are transverse waves.

Like other waves, such as water waves and waves on a rope, electromagnetic waves carry energy from one place to another. But unlike other waves, electromagnetic waves do not carry energy by causing matter to vibrate. It is the electric and magnetic fields that vibrate. This explains why electromagnetic waves can travel in a vacuum (where there is no matter). But it does not mean that electromagnetic waves cannot travel through a medium. They certainly can. Light, for example, can be transmitted with a medium—as through the atmosphere—or without a medium—as through space.

Figure 3–1 *The colors of light produced by fireworks are one familiar form of electromagnetic waves. What type of wave is light?*

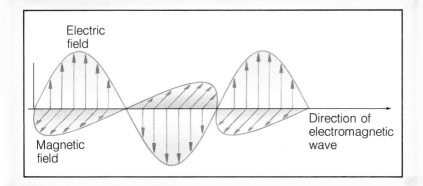

Figure 3-2 *In a transverse wave, the direction of the wave energy is at right angles to both the electric and the magnetic fields.*

Production and Transmission of Electromagnetic Waves

When you first started studying waves, you learned that the source of any wave is a vibration. For example, a vibrating bell causes air particles to move back and forth, producing a sound wave. This same idea about a vibrating source is true for electromagnetic waves as well. However, rather than a source setting up vibrations in a medium, the source of an electromagnetic wave sets up vibrating electric and magnetic fields.

To understand how electromagnetic waves are produced, you must first become familiar with the **atom.** Atoms are the building blocks of matter. An atom consists of a central core, or nucleus, surrounded by tiny particles called electrons. Electrons do not have set positions. Instead, they constantly move about the nucleus.

Electrons are charged particles that can produce electric and magnetic fields. But in order to create the vibrating electric and magnetic fields that are characteristics of an electromagnetic wave, electrons must move. A charged particle, such as an electron, moving back and forth creates electric and magnetic fields that move back and forth, or vibrate. **The source of all electromagnetic waves is charge that is changing speed or direction.** Visible light, for example, is produced by electrons jumping between different positions in an atom. According to modern theories used to describe atoms, electrons move at different distances from the nucleus according to the amount of energy they have. But an electron can absorb more energy and thereby move to another

ACTIVITY

THINKING

Electromagnetic Waves in Your Life

1. Look around your home at the devices and appliances you use every day. Name four objects in your home that produce electromagnetic waves.

2. Describe the type of electromagnetic wave that each object produces.

What type of electromagnetic wave is most common on your list?

Do electromagnetic waves play an important role in your life?

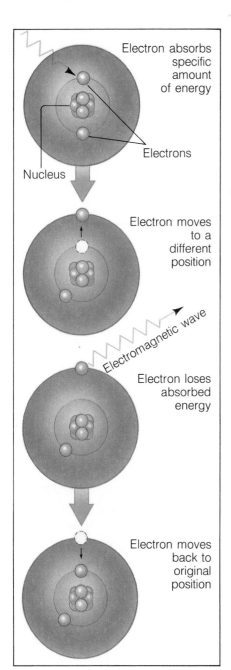

Electron absorbs specific amount of energy

Electrons

Nucleus

Electron moves to a different position

Electromagnetic wave

Electron loses absorbed energy

Electron moves back to original position

Figure 3-3 *All forms of electromagnetic waves have their source in the atom. In particular, visible light is produced when an excited electron returns to its normal position, releasing energy in the form of an electromagnetic wave.*

position. This move, however, is not a stable one. That is, the electron will not remain in its new position for long. Eventually, it will lose the extra energy and fall back to its original position. As it falls back, it creates vibrating electric and magnetic fields. These fields form the electromagnetic waves that carry the released energy.

Other types of electromagnetic waves are also created in atoms. For example, electrons moving back and forth in an antenna create radio waves. X-rays are produced when electrons slow down abruptly as they collide with a target in an X-ray tube. Gamma rays are produced when the nucleus of an atom gives up extra energy.

The speed of all electromagnetic waves is the same—300 million meters per second in a vacuum. This speed is usually referred to as the speed of light. The speed is slightly slower in air, glass, and any other material. To appreciate just how great this speed is, consider the following: Light from the sun travels 150 million kilometers to Earth in about 8 minutes! Nothing known in the universe travels faster than the speed of light.

Figure 3-4 *Electromagnetic waves carry light from distant stars. If light required a medium for transmission, it would not be able to travel through space to Earth.*

3–1 Section Review

1. What is an electromagnetic wave?
2. What are the physical characteristics of an electromagnetic wave?
3. What is an electric or magnetic field?
4. How is visible light produced?

Critical Thinking—*Relating Concepts*
5. How are sound and light alike? How are they different?

PROBLEM ??? Solving

Reaching for the Stars

Suppose you have been given the following assignment: Determine the composition of three distant stars. How can you possibly do it? After all, you cannot travel billions of kilometers to get a sample of each. And, what's more, all you have been given is a few strips of paper with what seem like some silly lines on them. Whatever will you do?

Luckily for you, those strips of paper are all you need. All elements produce a characteristic set of lines when they are heated, and the light given off is passed through a device known as a spectroscope. The lines are called spectral lines. Every element has its own set of spectral lines—much like a fingerprint.

Examine the spectral lines of the following elements. Compare them with the spectral lines labeled A, B, and C. Determine which elements are in the stars that produced A, B, and C.

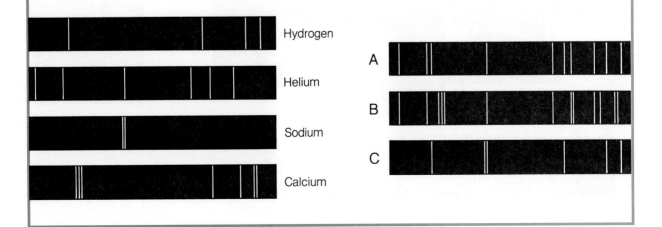

Hydrogen

Helium

Sodium

Calcium

A

B

C

Guide for Reading

*Focus on these questions as
you read.*

▶ *How is the electromagnetic
spectrum organized?*

▶ *What types of waves make
up the electromagnetic
spectrum?*

3-2 The Electromagnetic Spectrum

Now that you know what electromagnetic waves are, you might be wondering how sunlight is different from X-rays if both are electromagnetic waves that travel at the same speed. Electromagnetic waves, like all types of waves, are described by their physical wave features: amplitude, wavelength, and frequency. And it is these characteristics that can vary and thereby produce many different kinds of electromagnetic waves.

Electromagnetic waves are often arranged in order of wavelength and frequency in what is known as the electromagnetic spectrum. Because all electromagnetic waves travel at the same speed, if the frequency

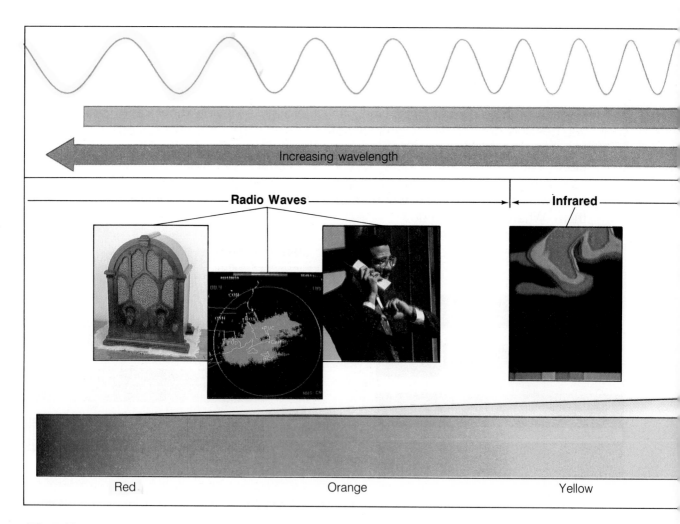

Increasing wavelength

Radio Waves ——————————— Infrared

Red   Orange   Yellow

of a wave changes, then the wavelength must change as well. (Remember: Speed = Wavelength × Frequency.) Waves with the longest wavelengths have the lowest frequencies. Waves with the shortest wavelengths have the highest frequencies. Look again at the formula for wave speed. Do you see why these statements must be true? The **electromagnetic spectrum** ranges from very long-wavelength, low-frequency radio waves to very short-wavelength, high-frequency gamma rays. The amount of energy carried by an electromagnetic wave increases with frequency. Figure 3–5 gives you a good idea of the nature of the electromagnetic spectrum. Notice that only one small region is visible. The rest of the spectrum is invisible!

The electromagnetic spectrum covers a tremendous range of frequencies and wavelengths. Comparing the frequencies of radio waves to those of visible

Figure 3–5 *Electromagnetic waves are arranged according to their increasing frequency and decreasing wavelength in the electromagnetic spectrum. Although the many applications of electromagnetic waves are very different, they all depend on the same basic type of waves.*

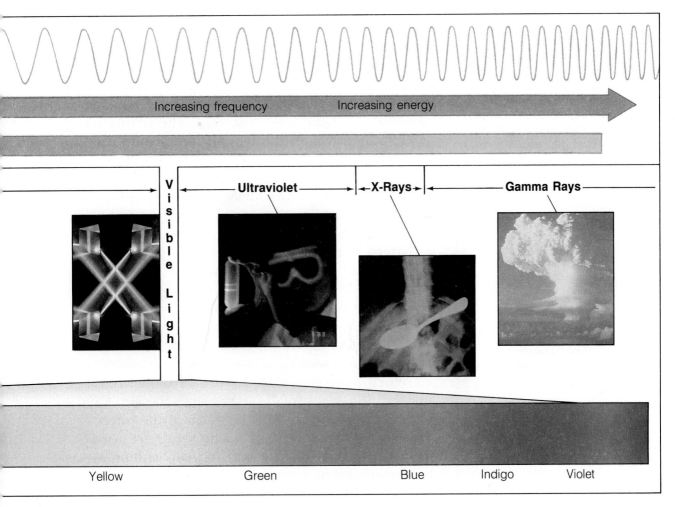

Increasing frequency   Increasing energy

Visible Light — Ultraviolet — X-Rays — Gamma Rays

Yellow   Green   Blue   Indigo   Violet

light is like comparing the thickness of this page to the Earth's distance from the sun! Because the differences are so great, the various regions of the spectrum are given different names.

Radio Waves

The waves in the electromagnetic spectrum that have the lowest frequencies and longest wavelengths are called **radio waves**. Radio waves are produced when charged particles move back and forth in instruments such as antennas. When you think of radio waves, do not confuse them with the waves you hear coming out of your radio. Those waves are sound waves. Radio waves are the waves used to transmit information from the antenna of a broadcasting station to the antenna on your radio or television.

Perhaps you are wondering how your favorite song or television show can be carried by a wave. The answer is rather interesting. When radio waves are transmitted, one of two characteristics of the wave can be varied—either the amplitude of the wave or the frequency of the wave. The variation in either amplitude or frequency of a wave is called **modulation** (mahj-uh-LAY-shuhn). The setting on your radio indicates the type of modulation used to carry the information to your radio: AM means amplitude modulation, and FM means frequency modulation. At the broadcasting station, information (music or words or pictures) is converted from its original form to an electronic signal and then to a pattern of changes in either the amplitude or frequency of a radio wave. When the radio wave is received by an antenna of a radio or television, the pattern is converted back to its original form. The sound portions of most television broadcasts are carried as AM waves while the picture portions are carried as FM waves.

When radio waves are sent out from a broadcasting station, they spread out through the air. Any antenna tuned in to the frequency of the waves receives the waves. Usually radio waves are used when a message is being sent to many antennas—as in a television or radio broadcast. However, radio waves can be disturbed by obstacles or weather

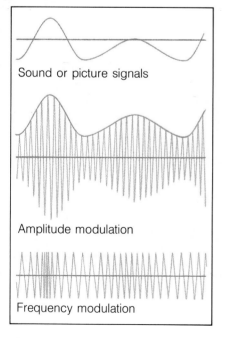

Figure 3–6 *Sounds at a broadcasting station are converted into electric sound signals. The sound signal is then used either to adjust the frequency (FM) or amplitude (AM) of a radio wave. Notice how the crest of the sound signal affects each of the radio waves.*

Sound or picture signals

Amplitude modulation

Frequency modulation

conditions. Suppose, for example, a small town is located deep in a valley. Many of the radio waves will be unable to pass through or around the surrounding mountains and the town will be unable to receive the signals. In situations such as this, it is important to protect the radio waves from being disturbed. One way that this can be accomplished is to send the radio waves through a cable wire that travels directly to the destination of the signal. This is basically how cable television works. Perhaps you have heard that cable television is often used in areas where television reception is poor. Cable television is, as its name suggests, television signals carried by radio waves through cables that protect the signal.

Radio waves are also used in medicine. Radio waves and strong magnetic fields are used to cause different atoms in the body to vibrate. By analyzing the response of different sections of the body to different frequencies, doctors can create pictures of parts of the human body, including the brain, without harming the cells. This procedure is known as magnetic resonance imaging.

In astronomy, radio waves have a different application. Observations of space are often blocked by conditions in Earth's atmosphere. Radio waves pass through the atmosphere unaffected, however. For this reason, astronomers have constructed telescopes that study radio waves from space. The main part of a radio telescope is shaped like a huge dish. The curved dish collects radio waves emitted from space

Figure 3–7 *Satellite dishes are used to record electromagnetic waves from space and form images of distant objects, such as this remnant of an exploding star, or supernova. Radio waves are also used to form images of structures within the human body. This image shows the hips and pelvic region.*

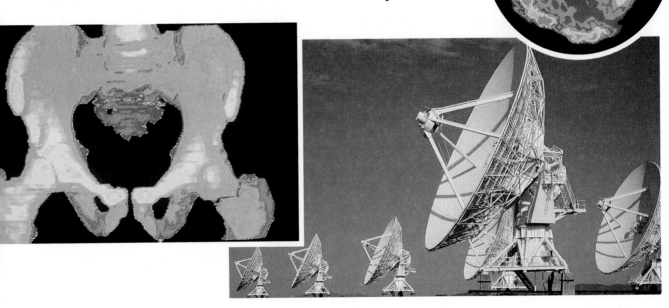

and focuses the waves toward an antenna. The radio waves are then fed to a computer, which processes the information and converts it into an image. Only in the past few decades have scientists learned that electromagnetic waves are radiated from space across the entire electromagnetic spectrum—not just the visible portion. So in addition to radio telescopes, devices for collecting information in every region of the spectrum have been developed.

MICROWAVES One particular group of radio waves is that of **microwaves.** Microwaves are the highest frequency radio waves. The wavelengths of microwaves are only a few centimeters long.

The application of microwaves that is probably most familiar to you is a microwave oven. One of the advantages of a microwave oven is that food can be heated in a short amount of time without heating the dish that the food is in. How can this happen? The answer is that microwaves pass right through some substances but are absorbed by others. Water and other molecules in the food absorb energy from the microwaves, causing the food to heat up. Glass and plastic containers do not absorb microwaves and therefore do not become hot in a microwave oven. Metals, however, absorb microwaves. In fact, metals can absorb enough electromagnetic energy from microwaves to create a current of electricity. That is why metal containers should never be used in a microwave oven.

An application of microwaves with which you may be less familiar is communication. Yet actually, transmitting information is the most common use of microwaves. For example, microwaves are used for communication in cellular, or portable, telephones. Microwaves transmit information efficiently because they are not easily blocked by structures such as mountains, trees, and buildings.

Microwaves are also used in weather forecasting. Microwaves can penetrate clouds of smoke, but they are spread out by water droplets. By observing what happens to microwaves sent out into the atmosphere, weather forecasters can locate storms.

RADAR Short-wavelength microwaves are used in **radar.** Radar, which stands for *ra*dio *d*etecting *a*nd *r*anging, is used to locate objects and monitor speed.

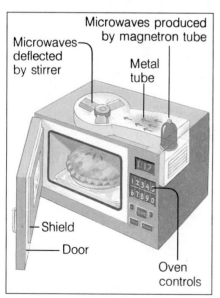

Figure 3–8 *Because microwaves are absorbed by most foods, they are used in cooking. Microwaves cause the molecules of the food to vibrate, which increases the temperature of the food. How does the speed of microwave cooking compare with conventional cooking methods?*

The procedure is the same as sonar for sound waves. A radar device operates by sending out short pulses of radio waves. Any object within a certain distance will reflect these waves. A receiver picks up the reflected waves, records the length of time it takes for the waves to return, and then calculates the distance to the object. Radar is used to monitor the positions of airplanes taking off and landing at airports and to locate ships at sea—especially in heavy fog. Radar is used to keep track of satellites that are orbiting the Earth. In addition, weather forecasters use radar to locate and track storms.

In Chapter 2, you learned about the Doppler effect, which is caused when either the source of a sound or the receiver is moving. The sound waves bunch up or spread out, changing the frequency (and thus the pitch) of the sound. The Doppler effect also exists for electromagnetic waves. It is what enables a police officer on the side of the road to determine the speed of a moving car. If a police officer sends microwaves to a moving car, the waves will be reflected. But because the car is moving, the rays coming back will have a different frequency from the rays sent out. The radar device uses the change in frequency to calculate the speed of the car.

Infrared Rays

Electromagnetic waves with frequencies slightly lower than visible light are called **infrared rays**. Although infrared rays cannot be seen, they can be felt as heat. You can feel infrared rays as heat from a light bulb, a stove, or the sun. Nearly 50 percent of the rays emitted by the sun are in the infrared region.

All objects give off infrared rays. The amount of infrared given off by an object depends on the temperature of the object. Warmer objects give off more infrared rays than colder objects do. Cool objects can absorb the energy from infrared waves and become hotter. For this reason, infrared lamps are used to keep food hot, relieve sore muscles, and dry paint and hair.

Infrared cameras are devices that take pictures using heat instead of light. Such cameras allow

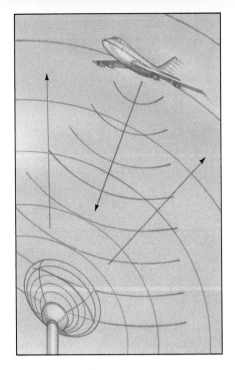

Figure 3-9 In a radar device, a radio transmitter sends out high-frequency waves that bounce off objects and return as echoes picked up by a receiver. Can you think of other important uses for radar?

Heat and Light

1. Place a thermometer on a dinner plate.

2. Position an incandescent light bulb 10 cm above the thermometer.

3. Turn on the bulb. After about 5 minutes, record the temperature.

4. Repeat the procedure, but this time use a fluorescent bulb.

Which light source operates at a higher temperature? Is this an advantage?

Figure 3-10 *Infrared light cannot be seen, but it can be detected as heat and used to produce a thermogram. Hotter areas appear white, and cooler areas appear black or blue. Note the dog's cold nose!*

pictures to be taken at night. This ability has special applications for researchers who wish to observe the habits of animals during the night, for security systems, and for military operations.

Certain types of photographs that are taken using heat are called thermograms. Thermograms can identify the warm and cold areas of an object or person. Thermograms can be used to detect heat loss in a building. They are also used in medicine. Sometimes, unhealthy tissue will become hotter than surrounding healthy tissue. Thermograms can be used to identify such tissue.

People studying pieces of art—either to learn from them or to determine their authenticity—make use of infrared rays. To the unaided eye, a painting on a canvas is all that appears on the surface. But sometimes there is more than meets the eye. During certain periods in history (and for various reasons), artists painted on used canvas—one painting right on top of another. A painting thought to be lost or missing may have been painted over. In other

Figure 3-11 *You probably think these photographs were taken on beautifully bright days. Quite the contrary. The photographs were taken in the darkness of night using an infrared camera.*

instances, it is discovered that an artist changed his or her style as the painting progressed. And in some cases, the original signature on a painting may be painted over with a more well-known one so that the owner can sell the painting for more money. How do art experts find out the truth? One technique involves shining infrared rays onto a painting. Infrared rays are able to penetrate the thin upper layers of an oil painting and reveal the layers beneath the surface. Other forms of electromagnetic energy—X-rays and ultraviolet rays—are also used for the same purpose.

The Visible Spectrum

The electromagnetic waves that you can see are called **visible light**. Notice that visible light—red, orange, yellow, green, blue, indigo, and violet—is only an extremely small portion of the spectrum. The rest of the electromagnetic spectrum is invisible to the unaided eye. This explains why you can be surrounded by electromagnetic waves every day but never see them!

Despite its small range, the visible spectrum is of great importance. Life on Earth could not exist without visible light. Nearly half the energy given off by the sun is in the form of visible light. Over billions of years, plants and animals have evolved in such ways as to be sensitive to the energy of the sun. Without such adaptations, life would be impossible. Visible light is essential for photosynthesis, the process by which green plants make food. Other forms of life eat either green plants or animals that have eaten green plants. Forms of energy taken from the sun by plants and microorganisms millions of years ago are locked up in the coal and oil used as energy resources today. You will learn more about visible light in the next section and in Chapter 4.

Ultraviolet Rays

Electromagnetic waves with frequencies just higher than visible light are called **ultraviolet rays**. The energy of ultraviolet rays is great enough to kill living cells. So ultraviolet lamps are often used to kill germs in hospitals and to destroy bacteria and preserve food in processing plants.

ACTIVITY READING

By Land or by Sea

Light has played an important role in history. One use of light has been as a signal in situations where sound could not be used. Read *Midnight Alarm: The Story of Paul Revere's Ride* by Mary K. Phelan and enjoy a story about the use of light in the American Revolution.

Figure 3–12 *Some scientific research combines more than one area of science. For example, Dr. Shirley Ann Jackson is a talented physicist who works in the field of optoelectronics. Her research involves designing electronic devices that produce, regulate, transmit, and detect electromagnetic radiation. Some of the applications of her research are improving the nature of communication throughout the world.*

Figure 3–13 *The sunflowers you are accustomed to seeing resemble the cheerful yellow flower shown on the right. A honeybee, however, sees something quite different. The photo on the left shows the sunflower seen in ultraviolet light.*

Figure 3–14 *When rocks containing fluorescent minerals are exposed to invisible ultraviolet light, they glow. Why might this property be important to geologists?*

Although ultraviolet light is invisible to humans and many other animals, it can be seen by many insects. Flowers that appear to be the same color in visible light are much different under the ultraviolet rays seen by insects. Do you remember the honeybee you read about at the beginning of this chapter? The honeybee took advantage of its ability to see into the ultraviolet region of the electromagnetic spectrum.

As you may already know, ultraviolet rays are present in sunlight. Although the eyes cannot see them, their effects can be felt as sunburn. When your body absorbs sunlight, ultraviolet rays cause your skin cells to produce vitamin D. Vitamin D is needed to make healthy bones and teeth. Despite this benefit, ultraviolet rays are harmful to body cells. (All electromagnetic waves beyond the visible region—ultraviolet, X-rays, and gamma rays—are harmful to living cells.) Tanning is the body's way of protecting itself against these harmful rays. But overexposure to ultraviolet rays can cause serious damage to the skin, eyes, and immune system.

Fortunately, a layer of ozone in the atmosphere absorbs most of the sun's damaging ultraviolet rays before they reach the Earth. Without this protective ozone layer, life on Earth would be impossible. Many scientists believe that this protective ozone layer is now being destroyed by chemicals released into the atmosphere by aerosol propellants, air-conditioning coolants, and various other sources.

X-Rays

Electromagnetic waves with frequencies just above ultraviolet rays are called **X-rays**. The energy of X-rays is great enough to pass easily through many materials, including your skin. Denser materials, however, absorb X-rays. Bone absorbs X-rays. When an X-ray picture of a part of your body is taken, the bones absorb the rays but the soft tissues do not. The picture that results shows the bones as white areas and the soft tissues as black areas.

Strong sources of X-rays have been detected deep in space. The sources are believed to be certain star formations. In addition to these star formations, exploding stars are known to give off most of their energy at the time of explosion in the form of X-rays.

Despite their usefulness in medical diagnosis, X-rays are a potential health hazard. Exposure of body cells and tissues to large amounts of X-rays over a lifetime can cause defects in cells. Lead absorbs almost all the X-rays that strike it. Can you think of an important use of lead, based on this property?

Gamma Rays

The electromagnetic waves with the highest frequencies and shortest wavelengths are called **gamma rays**. Gamma rays have the highest energy of the electromagnetic spectrum. Certain radioactive materials and nuclear reactions emit gamma rays.

Gamma rays have tremendous penetrating ability—even greater than that of X-rays. The energy of gamma rays is so great that they can penetrate up to 3 meters of concrete! Excessive exposure to gamma rays can cause severe illness.

Gamma rays have positive applications in medicine. A patient under observation can be injected with a fluid that emits gamma rays. A camera that

Figure 3–15 *Doctors can observe the result of a skier's unfortunate accident—a broken leg—by taking an X-ray. Similarly, doctors can use images formed by a gamma camera to study the skeleton of a healthy person.*

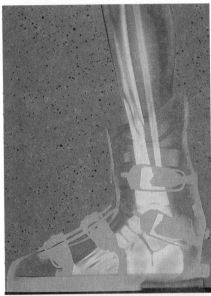

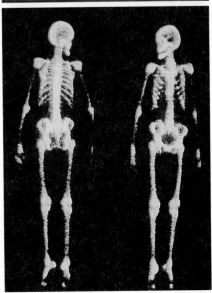

detects gamma rays—a gamma camera—can then be moved around the patient to form an image of the inside of the body according to how the rays are given off.

ACTIVITY WRITING

X-Ray Vision

Wilhelm Roentgen discovered X-rays in 1896. Within a few months, X-rays were being used in hospitals. Using books and other reference materials in the library, find out about Roentgen and his work with X-rays. Write a report on your findings. Include information on the benefits and dangers of X-rays.

Activity Bank

A Sunsational Experiment, p.144

1–2 Section Review

1. What is the electromagnetic spectrum? List the kinds of waves that make it up.
2. Moving along the spectrum from radio waves to gamma rays, what happens to frequency? Wavelength? Energy?
3. What is the difference between AM radio waves and FM radio waves?
4. Describe a use of each of the following: infrared rays, ultraviolet rays, X-rays, and gamma rays.

Connection—*Life Science*

5. Why is the destruction of the ozone layer dangerous? Design a poster to express the importance of protecting the ozone layer.

Guide for Reading

Focus on this question as you read.

▶ *How do luminous objects produce light?*

3–3 Visible Light

You are not capable of giving off light. But a firefly is. Have you ever seen a firefly light up? Certain organisms and objects give off their own light. Anything that can give off its own light is called a **luminous object**. The sun and other stars, light bulbs, candles, campfires, and fireflies are luminous objects.

Other objects are lit up, but not by their own light. When you stand in a spotlight, you are lit up by a luminous object (the spotlight). You can be seen because the light given off by a luminous object bounces off you. An object that can be seen because it is lit up is called an **illuminated object**. You see the moon because sunlight is reflected off its surface. The moon does not give off its own light. How would you describe the pages of this textbook? The lamp in your room?

Production of Light

There are three ways in which a luminous object can be made to give off energy in the form of light. These three ways determine the type of light produced. **A luminous object can produce incandescent light, fluorescent light, or neon light.**

INCANDESCENT LIGHT   Have you ever seen the coils of a toaster oven heat up until they were red hot? Or the coals of a barbecue turn bright orange? Certain objects can be heated until they glow, or give off light. Light that is produced from heat is called **incandescent light**. An object that gives off incandescent light is said to be incandescent.

Have you ever touched a light bulb after it has been lit for a while? It is quite hot to the touch. Ordinary light bulbs in your home are incandescent. They produce light when electricity is applied to them. Inside the glass bulb of a light bulb is a thin wire filament made of the metal tungsten. Tungsten can be heated to over 2000°C without melting. When the light is switched on, electrons flow through the tungsten wire. Because the filament is thin, there is resistance to the electron flow. Electric resistance produces heat. Enough heat will cause the tungsten to glow, creating visible light.

FLUORESCENT LIGHT   You have probably seen long and narrow or circular white lights in your school, an office, or a department store. These lights are called **fluorescent lights**. Fluorescent light is cooler and uses much less electricity than incandescent light. Instead of being used to build up heat, electrons in fluorescent lights are used to bombard molecules of gas kept at low pressure in a tube.

Normally, you cannot see ultraviolet light. The inside of a fluorescent light, however, is coated with special substances called phosphors. Phosphors absorb ultraviolet energy and begin to glow, producing visible light. The color that a fluorescent bulb produces depends on the phosphors used.

Phosphors are sometimes added to laundry detergents to make white clothes appear whiter. Ultraviolet light in sunlight will cause the phosphors to glow, making the color appear brighter. Have you ever seen your clothes glow in the haunted house or fun

Figure 3–16 *A firefly is a luminous source of light. Although Jupiter and its moons shine brightly, they are not luminous objects. What type of objects are they? Why can we see them?*

Figure 3–17 *Light bulbs and hot coals are examples of objects that produce light as they are heated. What are these types of luminous objects called?*

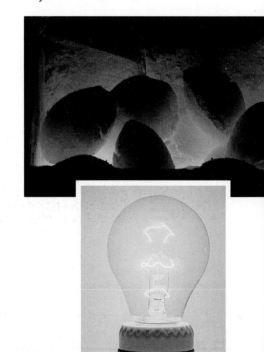

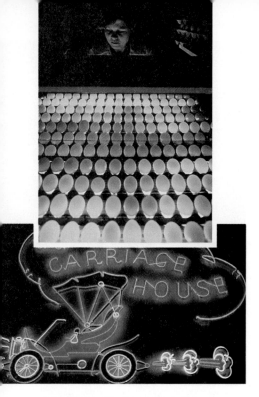

Figure 3–18 *Fluorescent lights have a number of industrial applications, such as illuminating eggs for inspection. Neon lights are usually more brightly colored and used for decoration or advertisement.*

Activity Bank

Mystery Message, p.145

Guide for Reading

Focus on this question as you read.

▶ *Is light a particle, a wave, or both?*

house at an amusement park? What kind of light do you think is used to make this happen?

NEON LIGHT   Perhaps you have seen thin glass tubes of brightly colored lights in an advertisement or sign. These lights were probably **neon lights**. Neon light is similar to fluorescent light in that it is cool light. When electrons pass through glass tubes filled with certain gases, light is produced. The most common type of gas used is neon gas. The light produced from neon gas is bright red. If other gases are added, however, different colors are produced.

Mercury vapor produces greenish-blue light that does not create much glare. So mercury vapor lamps are used to light streets and highways. Sodium vapor lamps, which give off a bright yellow-orange light, use less electricity than mercury vapor lamps. In many locations, sodium vapor lamps are replacing mercury vapor lamps.

3–3 Section Review

1. What is a luminous object? An illuminated object?
2. How does an incandescent bulb produce light?
3. How does a fluorescent source produce light?
4. How is fluorescent light similar to neon light?

Critical Thinking—*Drawing Conclusions*
5. Why are fluorescent lights helpful in preserving Earth's natural resources?

3–4 Wave or Particle?

Throughout this chapter you have read about light as a wave and about the properties of electro-magnetic waves. The wave model of light, which has been the prevailing theory since the early 1800s, successfully explains most of the properties and behavior of light. In the early 1900s, however, scientists discovered something unusual about light—something that made them modify the wave theory.

Scientists shone violet light onto the surface of certain metals. The energy carried by the light was absorbed by electrons in the atoms of the metal

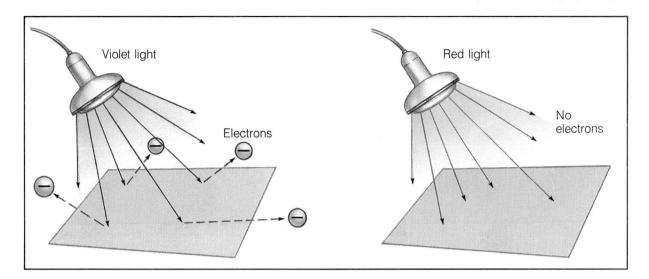

Figure 3–19 *The energy of individual photons of violet light can produce an electric current. The energy of individual photons of red light cannot. What theory of light explains this effect?*

plate. This energy knocked electrons out of some of the atoms in the metal plate. In fact, enough electrons were knocked off the metal plate to cause an electric current to flow.

The scientists then repeated the experiment with red light. To their surprise, nothing happened! No matter how long the red light was shone or how bright it was, no electrons were ever knocked out of the metal's atoms. The dimmest violet light produced a current, but the strongest red light did not!

As you can imagine, this caused confusion. According to the wave theory, if red light strikes a metal as a continuous wave, eventually the electrons should "soak up" enough energy to escape from their atoms. But that does not happen.

Suppose, though, that light acts more like a stream of particles than like a wave. Each individual particle would be a tiny bundle of energy related to the amount of energy absorbed by an electron. Each individual particle, or **photon**, of red light acting on its own could never knock an electron from its atom. Each photon must have a different amount of energy for each color of light. However, no single red light photon contains enough energy to do the job, no matter how long the light is on or how bright it is. On the other hand, violet light photons carry more energy than red light photons. So a single violet light photon can impart enough energy to an electron to knock it right out of its atom.

Further experimenting showed that photon energy increases as frequency increases and wavelength decreases. Thus, photon energy increases across the electromagnetic spectrum from radio

Figure 3–20 *Solar cells, such as these on a satellite in orbit around the Earth, use the energy of sunlight to produce a useful electric current. Why are solar cells becoming increasingly important to energy conservation and ecology?*

Figure 3–21 *When two beams of light from two slide projectors intersect, they pass through each other without colliding. The images produced on the screen are clear. What theory of light explains this behavior?*

waves to gamma rays. Because the experiments involved both electrons and photons, the result came to be known as the **photoelectric effect.**

The photoelectric effect can only be explained by a particle theory of light. But many other properties of light can only be explained by a wave theory of light. Confused? Don't be. **Scientists today describe light and other electromagnetic waves as both particlelike and wavelike.** Although it may be difficult for you to picture both a particle and a wave at the same time, both models are necessary to explain all the properties of light. And this problem provides a good opportunity for you to remember that science is a way of explaining observations; it is not absolute knowledge.

ACTIVITY

DISCOVERING

Light: Particle or Wave?

1. CAUTION: *Do this step outdoors.* Turn on two water hoses. Aim the stream of water from one hose across the path of the stream of water from the other hose. Observe what happens to the stream of water from the second hose.

2. Darken a room and project a slide from a slide projector on the wall. Shine a flashlight beam across the projector beam in the same way you did the stream of water in step 1. Observe any effect it has on the projected picture.

Do the two streams of water particles act in the same way that the two beams of light do? Explain your answer.

■ Does this activity support a wave theory or a particle theory of light? Explain your answer.

3–4 Section Review

1. Why is light said to have two natures—wave and particle?
2. What convinced scientists of the particle nature of light?
3. Which carry more energy, ultraviolet photons or photons in gamma rays? Explain.

Connection—*You and Your World*
4. Many security systems are composed of a beam of light shining on a metal plate. As the electrons are knocked out of the plate, they create a current of electricity. If something crosses the beam of light, the flow of electricity stops and an alarm sounds. Why is the photoelectric effect necessary to explain this type of alarm?

The World at Your Fingertips

Click. The television goes on. Click again and the volume goes up. Click once more and the channel changes. Remote control, or controlling a device from a distance, has become a standard feature of many household products, ranging from television sets to garage-door openers. Beyond the household, remote control is used to control faraway devices such as satellites and guided missiles. Remote control makes many devices easier to operate—with just the push of a button. It also helps to perform tasks that would otherwise be difficult, dangerous, or even impossible.

There are different kinds of remote-control mechanisms. Consider the television again. When you push a button on the remote control, you set a tiny device into vibration. The vibration produces an infrared beam carrying a signal that is sent to the television. The signal depends on the button you push. A detector on the television receives the infrared beam signal and converts it to an electronic signal. The electronic signal controls several switches—on/off, channel selector, and volume. The switch that is activated by the electronic signal depends on which button you pressed. Infrared waves are also used to carry signals to VCRs and cable television boxes.

Other remote controls use radio waves. For example, radio waves are used to drive toy cars, fly model airplanes, and open garage doors. A remote-control device sends radio waves carrying signals to the object. A receiver decodes the signal and passes on the message to tiny motors controlling the object's movements. Some radio-control systems work with the aid of computers. Controllers on the ground use the combination of radio signals and computers to position antennas and operate other equipment on artificial satellites in orbit.

The next time you effortlessly push a remote-control button, think about the electromagnetic energy involved. From toy cars to guided missiles, electromagnetic waves really do put the world at your fingertips!

Laboratory Investigation

In the Heat of the Light

Problem

How do the effects of the various regions of the electromagnetic spectrum differ?

Materials *(per group)*

scissors
light bulb
convex lens
cardboard shoe box
prism
black paint
sensitive thermometer

Procedure 🔬 🔲 📷

1. Place the convex lens in front of the light bulb so that the light focuses through a small hole cut in the side of the shoe box. See the accompanying diagram.

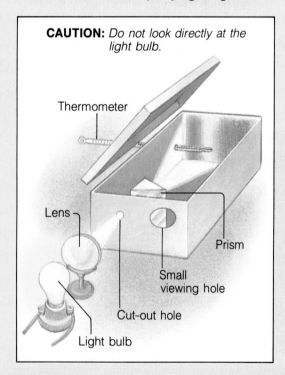

CAUTION: *Do not look directly at the light bulb.*

Thermometer

Lens

Prism

Small viewing hole

Cut-out hole

Light bulb

2. Position the prism in front of the hole in the box. A prism separates light into various wavelengths.
3. Use the black paint to paint the thermometer bulb so that it will absorb all the light energy that strikes it.
4. Place the thermometer bulb in the corner of the box. Allow a few minutes for the thermometer to register the temperature. Record the temperature.
5. Repeat Step 4 at several points, moving the thermometer bulb over slightly each time. Be sure to place the thermometer at points beyond the visible spectrum.

Observations

1. Record the temperatures in a data table.
2. Starting at the red end of the spectrum and moving toward the violet end, explain what happens to the temperature.

Analysis and Conclusions

1. Where is the warmest part of the spectrum?
2. Did you detect infrared or ultraviolet waves? How do you know?
3. How is the change in temperature related to the amount of energy carried by each wave?
4. If your temperatures did not vary in a regular pattern, what reasons can you give for the observation?
5. **On Your Own** Repeat the experiment with a bottle of water (or other fluid) outside the box and in front of the hole. How does this change your results?

Summarizing Key Concepts

3–1 Electromagnetic Waves

▲ An electromagnetic wave consists of an electric field and a magnetic field at right angles to each other and to the direction of motion of the wave.

▲ Electromagnetic waves can travel through a vacuum because they do not require matter to exist.

▲ Electromagnetic waves are produced by charge that is changing direction or speed.

▲ All electromagnetic waves travel at the same speed in a vacuum—300 million m/sec.

3–2 The Electromagnetic Spectrum

▲ The electromagnetic spectrum is an arrangement of electromagnetic waves according to wavelength and frequency.

▲ The electromagnetic spectrum includes radio waves, infrared rays, visible light, ultraviolet rays, X-rays, and gamma rays.

3–3 Visible Light

▲ Luminous objects give off light. Illuminated objects are lit up because light bounces off them.

▲ Incandescent light is produced from heat.

▲ Fluorescent light is produced when electrons are used to bombard gas molecules contained at low pressure.

▲ Light that is produced when electrons pass through glass tubes filled with gas is called neon light.

3–4 Wave or Particle?

▲ Visible light and other electromagnetic waves have a particlelike nature as well as a wave nature.

▲ The fact that light rays can knock electrons out of their atoms is referred to as the photoelectric effect.

Reviewing Key Terms

Define each term in a complete sentence.

3–1 Electromagnetic Waves
atom

3–2 The Electromagnetic Spectrum
electromagnetic spectrum
radio wave
modulation
microwave
radar
infrared ray
visible light
ultraviolet ray
X-ray
gamma ray

3–3 Visible Light
luminous object
illuminated object
incandescent light
fluorescent light
neon light

3–4 Wave or Particle?
photon
photoelectric effect

Chapter Review

Content Review

Multiple Choice

Choose the letter of the answer that best completes each statement.

1. Which of the following does not belong?
   a. X-rays
   b. sound waves
   c. gamma rays
   d. radio waves
2. The source of all electromagnetic waves is the
   a. air.
   b. sun.
   c. Earth.
   d. atom.
3. The electromagnetic spectrum arranges waves in order of
   a. frequency and wavelength.
   b. the alphabet.
   c. discovery.
   d. use and applications.
4. Microwaves are a type of
   a. X-ray.
   b. gamma ray.
   c. ultraviolet light.
   d. radio wave.
5. Thermograms are produced using
   a. infrared rays.
   b. gamma rays.
   c. X-rays.
   d. visible light.
6. The waves with the highest energy in the electromagnetic spectrum are
   a. gamma rays.
   b. radio waves.
   c. ultraviolet rays.
   d. visible light.
7. Light produced from heat is called
   a. neon light.
   b. fluorescent light.
   c. incandescent light.
   d. illuminated light.
8. Substances that glow when exposed to ultraviolet light are called
   a. phosphors.
   b. photons.
   c. photoelectric.
   d. neon.

True or False

If the statement is true, write "true." If it is false, change the underlined word or words to make the statement true.

1. Electromagnetic waves are <u>longitudinal</u> waves.
2. The <u>invisible</u> spectrum contains all the colors of the rainbow.
3. <u>Ultraviolet</u> rays can be felt as heat.
4. Waves with frequencies just above the ultraviolet region are called <u>gamma</u> rays.
5. Objects that can be seen because light bounces off them are called <u>luminous</u>.
6. In fluorescent lights, phosphors glow in response to <u>infrared</u> rays.
7. A particle of <u>light</u> is called a <u>photon</u>.
8. The movement of electrons because of energy absorbed from photons of light is called the <u>Doppler</u> effect.

Concept Mapping

Complete the following concept map for Section 3–2. Refer to pages R6–R7 to construct a concept map for the entire chapter.

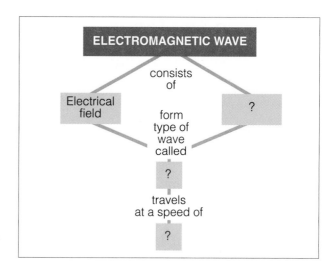

Concept Mastery

Discuss each of the following in a brief paragraph.

1. How is visible light produced?
2. How is the electromagnetic spectrum organized in terms of frequency, wavelength, and photon energy?
3. What happens to the wavelength of an electromagnetic wave if its frequency increases? Why is this true for all electromagnetic waves?
4. Are the wavelengths of radio and television signals longer or shorter than those detectable by the human eye? Describe the type of waves that carry these signals.
5. Explain why the wave theory of light cannot explain the photoelectric effect.
6. A person lost in the forest at night may signal for help by turning a flashlight on and off according to a code known as Morse code. This is actually a modulated electromagnetic wave. Is it AM or FM? Explain.
7. List five examples of luminous objects.
8. How is light produced in an incandescent light bulb? In a fluorescent light bulb? In a neon light?

Critical Thinking and Problem Solving

Use the skills you have developed in this chapter to answer each of the following.

1. **Applying concepts** How is it that someone in the United States listening on the radio to a music concert taking place in London, England, can hear the music before someone sitting in the audience can?
2. **Making calculations**
   a. A police radar signal has a frequency of about 11 billion Hz. What is the wavelength of this signal?
   b. What is the frequency of a microwave that has a wavelength of 1.50 cm?
3. **Drawing conclusions** What important information can be gathered from a thermogram of a house?
4. **Applying concepts** Suppose you are building an incubator and you need a source of heat. Would you use an incandescent or fluorescent light bulb? Explain your answer.
5. **Recognizing relationships** Describe the relationship between ultraviolet waves and the ozone layer.
6. **Making comparisons** Explain how a row of billiard balls rolling across a table in single file illustrates the particle nature of light.
7. **Using the writing process** Develop an advertising campaign praising the merits of fluorescent lights.

Light and Its Uses

Guide for Reading

After you read the following sections, you will be able to

4-1 Ray Model of Light
- Describe the nature of a light ray.

4-2 Reflection of Light
- Compare regular and diffuse reflections.

4-3 Refraction of Light
- Describe the process of refraction.

4-4 Color
- Account for the color of opaque and transparent objects.

4-5 How You See
- Explain how you see.

4-6 Optical Instruments
- Describe the operation and uses of several optical instruments.

4-7 Lasers
- Explain how a laser works.

Imagine a sturdy glass disc, smaller than a record album, that can hold enough information so that if it were printed and stretched out it would span over 5 kilometers! And imagine that the information on the disc could be used over and over without scratching or wearing out the disc. Well, you don't really have to imagine. This amazing disc is already in use. It is called a compact disc, or CD. You have probably even seen the pretty colors of a compact disc shimmering in the light.

Besides their small size, CDs have many advantages over vinyl record albums. CDs produce extremely clear sound with no background noise. And information on a CD is found with a push of a button. How can all this be accomplished?

The answer is through the use of light. Instead of sharp instruments such as needles, CDs depend on beams of light. A beam of light creates a code on the disc, and a beam of light reads the code when it is popped into a compact-disc player.

Sound strange? In this chapter you will learn about light—how it bounces and bends, how colors are produced, and how light is used in a great variety of instruments and devices.

Journal *Activity*

You and Your World Have you ever stopped to admire a beautiful rainbow? Describe a rainbow in your journal. Include the conditions that formed the rainbow you saw. Then write a short poem about rainbows.

Spinning compact discs are a shining example of the versatility of light.

ctivity Bank

The Straight and Narrow, p.146

4–1 Ray Model of Light

Have you ever been frightened by your own shadow? Or made shadow figures on a wall with your hands? Shadows occur due to the fact that light travels in straight lines. This is not difficult to understand. If you turn on a flashlight in a dark room, you see a straight beam of light. If an object gets into that beam, the object blocks some of the light and a shadow is created. The light does not bend around the object.

Shadows are quite common. Even the Earth and the moon create shadows with the sun's light. An eclipse occurs when either the Earth or the moon passes through the other's shadow.

Notice that neither the wave nature nor the particle nature of light is necessary to discuss shadows—only the assumption that light travels in straight lines. This assumption has led to the ray model of

Figure 4–1 *According to the ray model of light, a luminous object emits light rays in all directions. The existence of shadows supports the ray model of light. Why do shadows form?*

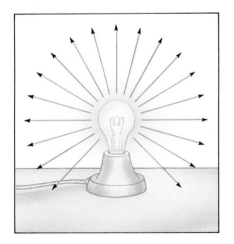

light. **The ray model of light assumes that light travels in straight-line paths called light rays.** Recall from Chapter 1 that straight arrows called rays were used to discuss such wave activities as reflection and refraction. Light rays are the same type of rays.

The ray model of light assumes that a luminous object sends out light rays that spread out in all directions in straight lines. An image is formed when the light rays leaving the object from the same point meet.

As you learn about light and its uses you will discover that different aspects and applications of light depend on either its particle nature or its wave nature. The ray model does not contradict the wave-like or particlelike characteristics of light. Instead, it puts aside the actual nature of light and concentrates on its behavior. In this way the ray model makes it easier for you to understand and describe activities of light such as reflection, refraction, and the formation of images.

4–1 Section Review

1. Describe the ray model of light.
2. What is a light ray?
3. When is an image formed?

Connection—*Mathematics*

4. What must happen to at least one of the light rays if two light rays that leave a point in different directions eventually meet?

4–2 Reflection of Light

When you bounce a ball on the pavement, it bounces back to you. If you bounce it at an angle, it bounces off the pavement at the same angle it bounced on the pavement. But what happens if you try to bounce the ball on a bumpy field or a rocky playground? The ball bounces off in a wild direction. You have no way of predicting where it will go.

Guide for Reading

Focus on these questions as you read.

▶ *What is the relationship between kinds of reflection and reflecting surfaces?*

▶ *How are mirrors classified according to shape?*

Light behaves in much the same way as the ball. When light strikes a surface, some of the light bounces back. (The rest is either absorbed by the material or is transmitted through it.) **The bouncing back of light is called reflection.** Recall what you learned about reflection in Chapter 1. Like the ball, light will be reflected off a surface at the same angle as it strikes the surface. Does this sound familiar to you? It should! This is the law of reflection. Like all waves, light obeys the law of reflection. This means that when light is reflected from a surface, the angle formed by the incoming, or incident, ray and the normal equals the angle formed by the outgoing, or reflected, ray and the normal. The normal is an imaginary line perpendicular to the surface.

Kinds of Reflection

Why can you see your reflection in a piece of glass but not in a wall? In both cases light is reflected off a surface. The answer lies in how the light is reflected. **The type of surface the light strikes determines the kind of reflection formed.**

A piece of glass has a smooth surface. All the rays reaching the glass hit it at the same angle. Thus, they are reflected at the same angle. This type of reflection is called **regular reflection.** It is similar to the ball bouncing off the pavement.

The surface of a wall is not really smooth. This may surprise you because you probably think that most walls have a smooth surface. If you were to magnify the surface of a wall, however, you would see that it is rough and uneven. If the surface of a wall were smooth, you would be able to see yourself in it! Sometimes surfaces are made to shine by coating them with materials that fill in the uneven spots. Many waxes and polishes make surfaces shinier by doing that.

Because the surface of a wall is not smooth, each light ray hits the surface at a different angle from the other light rays. Each ray still obeys the law

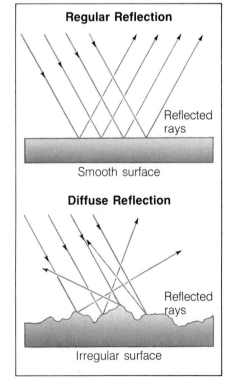

Regular Reflection

Reflected rays

Smooth surface

Diffuse Reflection

Reflected rays

Irregular surface

Figure 4–2 *Reflection from a smooth surface, or regular reflection, does not produce much scattering of light rays. Reflection from an irregular surface, or diffuse reflection, produces considerable scattering of light rays.*

Figure 4–3 *At first glance, it appears that there are two birds standing toe to toe. But actually there is only one bird. Reflection off the smooth surface of the water creates a clear image of the bird. What kind of reflection occurs when sunlight hits this grumpy-looking gorilla?*

of reflection, so each is reflected at a different angle from the others. Thus the reflected rays are scattered in all directions. Reflected light that is scattered in many different directions due to an irregular surface is called **diffuse reflection.** It is similar to the ball bouncing off the uneven field.

Although diffuse reflections are not desirable for seeing your image, they are important. If the sun's rays were not scattered by reflecting off uneven surfaces and dust particles in the air, you would see only those objects that are in direct sunlight. Anything in the shade of trees and buildings would be in darkness. In addition, the glare of the sunlight would be so strong that you would have difficulty seeing.

Reflection and Mirrors

The most common surface from which light is reflected is a mirror. Most mirrors are pieces of glass with a silver coating applied to one side. Glass provides an extremely smooth surface through which light can pass easily. Silver, which is not usually smooth, reflects almost all the light that hits it. Applying silver to the smooth surface of glass makes a very smooth reflecting surface.

ACTIVITY

DOING

Mirror, Mirror on the Wall

1. Stand 1 or 2 meters away from a full-length mirror. Have a friend or classmate place pieces of masking tape at the points on the mirror where you see the top of your head and your feet.

2. Compare the distance between the pieces of tape with your height.

3. Without changing the position of the mirror, move so that you are a distance of 4 or more meters from it. What happens to your image?

Derive a relationship between your height and distance from the mirror and the size and location of your image.

Figure 4–4 *Have you ever looked into a kaleidoscope? The magic of a kaleidoscope is created by a particular arrangement of mirrors.*

Figure 4–5 *This diagram shows how an image is formed by a plane mirror. Notice that the reflected rays do not actually meet. The brain, however, perceives them as having come from the point at which they would have met had they been straight.*

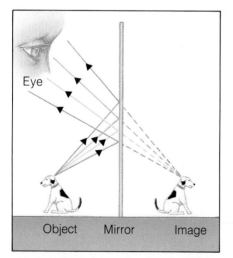

Any smooth surface that reflects light and forms images can be used as a mirror. The surface of a mirror can be perfectly flat or it can be curved. **Based on the shape of its surface, a mirror is classified as plane, concave, or convex.**

PLANE MIRRORS   A mirror with a perfectly flat surface is a **plane mirror.** An ordinary wall or pocket mirror is a plane mirror. Think about the image you see when you look into a mirror. Figure 4–5 shows how an image is formed. As you look at the figure, you should notice two things about the image. It is right side up, and the same size as the original object. And, as you know from experience, left and right are reversed. If you raise your left hand, the right hand of your image appears to be raised.

Where is the image when you look into a plane mirror? The image appears to be on the other side of the mirror—where you could almost reach forward and touch it. But you know this is not really so. Your brain is playing a trick on you. The human brain always assumes that light rays reach the eyes in a straight line. So even if the rays are reflected or bent, the eye records them as though they had traveled in a straight line. To better understand this, consider the bouncing ball again. If you did not know that the ball had bounced, it would seem as if the ball had come straight up through the pavement. This is what happens in a mirror. The brain traces light rays back to where they would have come from if they had been straight. They appear to have come straight through the mirror.

The light rays from an object reflected in a plane mirror do not actually meet. But the brain interprets an image at the point at which the light rays would have met had they been straight. This type of image is called a virtual image. As used here, the word virtual means not real. A virtual image only seems to be where it is. In other words, it can be seen only in the mirror.

CONCAVE MIRRORS   A mirror can be curved instead of flat. If the surface of a mirror curves inward, the mirror is called a **concave mirror.** You can experiment with a concave mirror by looking at the inner surface of a shiny metal spoon. Move the spoon back and forth and observe what happens to your image.

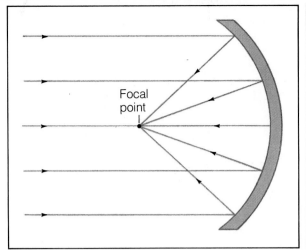 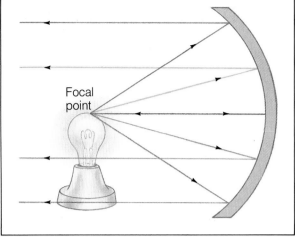

Figure 4-6 *Parallel light rays reaching a concave mirror are all reflected to the same point, the focal point. Light rays coming from the focal point, on the other hand, are reflected parallel to one another.*

Light rays coming parallel into a concave mirror are all reflected through the same point in front of the mirror. For this reason, concave mirrors are used in reflecting telescopes to gather light from space. The point in front of the mirror where the reflected rays meet is called the **focal point.** Figure 4-6 shows the focal point of a concave mirror. The distance between the center of the mirror and the focal point is called the focal length. Images formed by concave mirrors depend on the location of the object with respect to the focal point.

Notice the paths of the light rays in Figure 4-6. Light rays follow the same paths in the opposite direction if they are coming from the focal point rather than from space. All the light rays are reflected back parallel to one another in a concentrated beam of light. If you open a flashlight, you will find a concave mirror behind the bulb. The bulb is placed at the focal point of the mirror so that the reflected light forms a powerful beam. Concave mirrors are placed behind car headlights to focus the light beam. Concave mirrors are also used in searchlights and in spotlights.

CONVEX MIRRORS If you turn a shiny spoon over, you will notice that the surface curves outward. This is a **convex mirror.** The surface of a convex mirror curves outward like the surface of a ball. Reflected rays spread out from the surface of a convex mirror, as you can see from Figure 4-8 on page 98. The

Figure 4-7 *The image produced by a concave mirror depends on the position of the object in relation to the focal point. Where must the object be placed to produce a virtual image?*

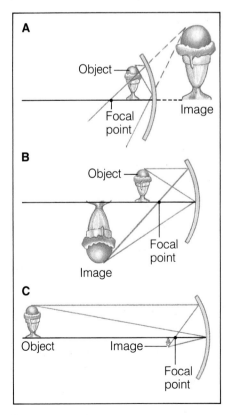

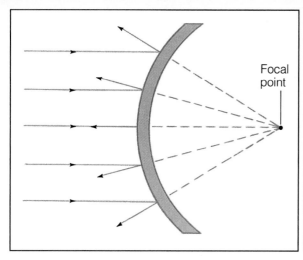

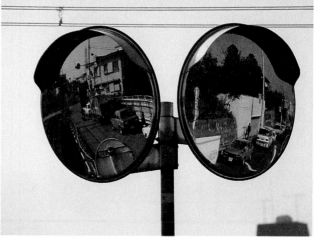

Figure 4-8 *Because a convex mirror spreads out reflected light rays from its very large area of reflection, it is often used to increase traffic visibility. What type of image does a convex mirror form?*

image formed by a convex mirror is always right side up and smaller than the object. Like the image formed by a plane mirror, the image appears behind the mirror. It is a virtual image.

Convex mirrors provide large areas of reflection. For this reason, they are used in automobile side-view and rear-view mirrors to obtain a wider view. They are also used in stores to provide security guards with a wide view of the shopping area. Convex mirrors, however, give a distorted indication of distance. Objects appear farther away than they actually are. Why is this an important concern when using a car mirror?

ACTIVITY

DOING

Reflection in a Spoon

1. Holding a metal spoon in front of you, observe your image on the back side of the spoon. Describe the image. What kind of mirror does this side represent?

2. Observe your image on the front side of the spoon. What kind of an image do you see? What kind of mirror does this side represent?

4-2 Section Review

1. How are regular and diffuse reflections related to the characteristics of the reflecting surface?
2. Describe the surface of and the images formed by a plane mirror. A concave mirror. A convex mirror.
3. What are some uses of each type of mirror?
4. A slide projector projects the image of a slide on a screen. Is this image real or virtual?

Connection—*Social Studies*
5. Archimedes is said to have burned the entire Roman fleet in the harbor by focusing the rays of the sun with a huge curved mirror. Explain if this is possible and what type of mirror it would have been.

4–3 Refraction of Light

You can win the stuffed animal at the carnival if you can throw a penny into the cup at the bottom of a fish bowl. Sounds easy, doesn't it? Actually, it is rather difficult—unless you know that the bending of light makes the cup appear to be where it is not!

You have just learned that light travels in straight lines. This means that it travels at a constant speed in a straight path through a medium. But light travels at different speeds in different mediums. So what happens when light passes from one medium to another? When light passes at an angle from one medium to another, it bends. (Do you remember learning in Chapter 1 that waves bend when they enter a new medium because part of the wave changes speed before the rest of the wave does?) **The bending of light due to a change in its speed is called refraction.**

When light passes from a less dense medium to a more dense medium, it slows down. This is the case when light passes from air to water. When light passes from a more dense medium to a less dense medium, it speeds up. This is the case when light passes from glass to air. To find out which way light bends, draw the normal to the boundary. If the wave speeds up, it bends away from the normal. If it slows down, it bends toward the normal. See Figure 4–9.

Because of refraction, a stick may look bent or broken when placed in a glass of water. If you are standing on the bank of a lake and you see a fish, it may appear closer to the surface than it actually is. Remember, the eyes see and the brain interprets light rays as if they were straight. The brain does not know that the rays were bent along the way.

Every medium has a specific **index of refraction,** which is a measure of the amount by which a material refracts light. **The index of refraction is the comparison of the speed of light in air with the speed of light in a certain material.** Because the

Guide for Reading

Focus on these questions as you read.

▶ *What is refraction?*

▶ *How do concave and convex lenses form images?*

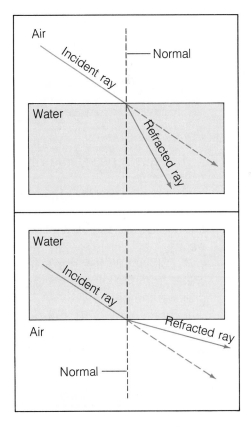

Figure 4–9 *As light passes from a less dense medium to a more dense medium, its speed decreases and it is refracted toward the normal (top). As light passes from a more dense medium to a less dense medium, its speed increases and it is refracted away from the normal (bottom).*

Figure 4–10 *The normal to this block of glass is horizontal across its length. When the light enters the glass, it bends up, toward the normal. When it leaves the glass on the other side, it bends down, away from the normal. Notice how the light breaks up as some of it is reflected.*

ACTIVITY

DOING

Penny in a Cup

1. Place a penny in a Styrofoam cup.

2. Fill a glass with water.

3. Position yourself so that the cup blocks the view of the penny.

4. Slowly fill the cup with water from the glass until you can see the penny.

Explain why you cannot see the penny without the water but you can see it with the water.

speed of light in air is always greater than in any other material, the index of refraction in any other material is always greater than one. The larger the index of refraction, the more light is bent.

Mirages

Perhaps you have had this experience: You were traveling in a car on a dry day when up ahead you saw a puddle of water shimmering on the roadway. But when you reached the spot where the puddle should have been, it was not there. And what's more, you saw another imaginary puddle farther ahead. The disappearing puddle is a type of mirage—not unlike the famous desert mirages sometimes seen in the movies.

A mirage is an example of refraction of light by the Earth's atmosphere. The index of refraction of air depends on the temperature of the air. When light passes from air at one temperature to air at another temperature, it bends. The greater the change in temperature, the greater the bending.

The puddle mirage exists because under certain conditions, such as a stretch of pavement or desert heated by intense sunshine, there are large temperature changes near the ground. Light from an object bends upward as it enters the hot air near the ground. It is refracted to the observer's eyes as if it had come from the ground or even from underground. So, for example, light from the sky can be refracted to look like a body of water.

Figure 4–11 *A small island with palm trees in the middle of a desert lake, right? No, wrong! This is only a mirage.*

Bending and Separating

Have you ever seen a beautiful rainbow? Or the colors of a rainbow as sunshine passed through a window, a drinking glass, or even a diamond? If so, have you wondered where rainbows come from?

You have learned that white light is made up of all the visible colors. Each color corresponds to a particular wavelength. If white light passes at an angle from air into another medium, its speed changes and it is refracted. Each wavelength is refracted by a different amount. Although the variation is small, it is enough to separate the different wavelengths in a beam of white light.

The longer the wavelength, the less bending there will be. Red, with the longest wavelength, is refracted the least. Violet, with the shortest wavelength, is refracted the most. The result is the separation of white light into the colors of the visible spectrum, which always appear in the same order—red, orange, yellow, green, blue, indigo, and violet. This process is called dispersion. If combined again, the various colors of the spectrum would form a beam of white light.

The piece of glass that forms the spectrum is called a prism. Notice that light bends as it enters the prism and as it leaves. The bending occurs as the light leaves the prism because the speed of light changes again as the light passes from glass back to air. Real rainbows are produced when tiny water droplets in the air act as prisms.

Figure 4–12 *The colorful rainbow seen above an irrigation device is produced by the dispersion of light. Because each wavelength of light is refracted a different amount in the droplets of water, sunlight is spread into an array of colors.*

Figure 4–13 *Passing white light through a prism separates the light into the various colors of the rainbow.*

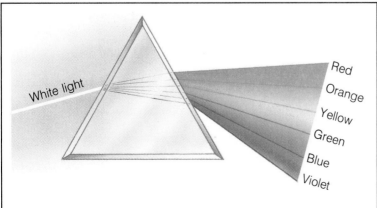

White light

Red
Orange
Yellow
Green
Blue
Violet

ACTIVITY READING

Refraction and Lenses

Have you ever used a magnifying glass, a camera, or a microscope? If so, you were using a **lens** to form an image. **A lens is any transparent material that refracts light.** The light is said to be focused through the lens. Most lenses are made of glass or plastic and have either one or two curved surfaces. As parallel rays of light pass through the lens, the rays of light are refracted so that they either come together or spread out. A lens that converges, or brings together, light rays is a **convex lens;** a lens that diverges, or spreads out, light rays is a **concave lens.**

CONVEX LENSES   A lens that is thicker in the center than at the edges is called a convex lens. As parallel rays of light pass through a convex lens, they are bent toward the center of the lens. The light rays converge. The point at which the light rays converge is the focal point.

Light is refracted as it enters a lens and again as it leaves the lens. The amount of refraction depends on the degree to which the lens is curved. A very curved lens will refract light more than a lens whose surface is only slightly curved.

Figure 4–14 *A convex lens converges light rays toward the center. The degree to which the lens is curved determines the amount of refraction. How does the focal length of a very curved lens compare with that of a slightly curved lens?*

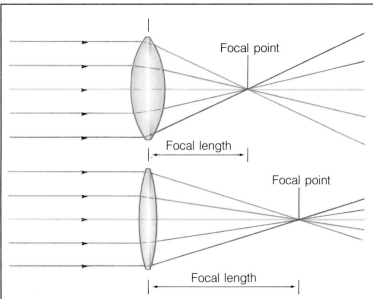

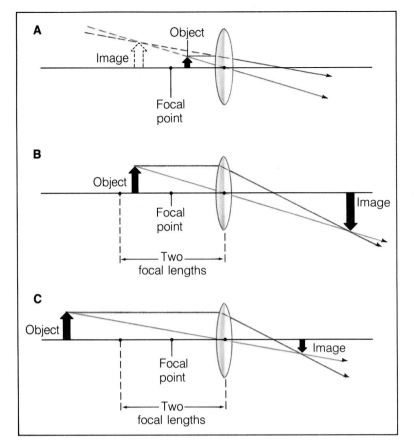

ACTIVITY

DOING

Firewater!

1. Place a candle 15 cm in front of a piece of clear glass or plastic.

2. Place a glass filled with water 15 cm behind the glass or plastic.

3. Put on safety goggles and then carefully light the candle. **CAUTION:** *Exercise care when working with matches and a lighted candle.*

4. Stand in front of the candle and hold a book or piece of cardboard between you and the candle so that your view of the candle is obscured. Look at the drinking glass. What do you see? Explain why you see what you do.

CONCAVE LENSES A lens that is thicker at the ends and thinner in the center is called a concave lens. As parallel rays of light pass through a concave lens, they are bent outward toward the ends of the lens. The light rays diverge.

Figure 4–16 *A concave lens diverges light rays toward the edges. The brain perceives the rays as if they were straight. How does the image compare with the object?*

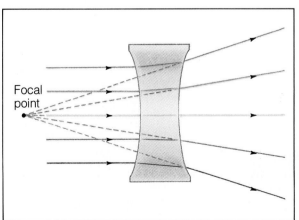
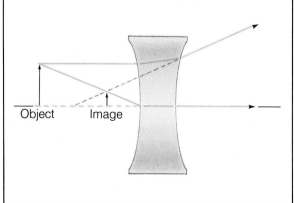

4-3 Section Review

1. Describe the process of refraction.
2. What is the index of refraction?
3. How can you determine which way a light ray will bend when it enters a new medium?
4. Which color of light is refracted the most? The least?

Critical Thinking—*Applying Concepts*

5. A magnifying glass is a simple lens. Explain which type of lens can be used as a magnifying glass and draw a diagram showing how light rays would be bent.

Guide for Reading

Focus on this question as you read.

▶ How is color related to the reflection, transmission, and absorption of light?

4-4 Color

You have just read that white light is broken into its individual colors by prisms. Yet you do not need a prism to see color. If all the colors of the rainbow are locked up within white light, how do objects have particular colors?

When Light Strikes

In order to understand why objects have color, you must know what happens when light strikes the surface of an object. **When light strikes any form of matter, the light can be transmitted, absorbed, or reflected.**

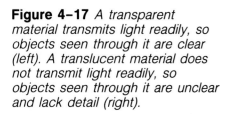

Figure 4-17 *A transparent material transmits light readily, so objects seen through it are clear (left). A translucent material does not transmit light readily, so objects seen through it are unclear and lack detail (right).*

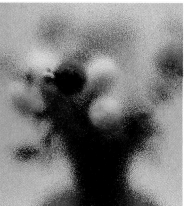

When light is transmitted, it passes through the substance it strikes. If the light is transmitted readily, the substance is said to be **transparent.** Objects seen through transparent substances are very clear. Glass, water, and air are transparent.

If light is transmitted through a substance that scatters the light, the image is unclear and lacks detail. A substance that transmits light but no detail is said to be **translucent.** Waxed paper and frosted glass are translucent substances. A translucent substance produces a fuzzy image when you look through it.

A substance that does not transmit light is said to be **opaque.** A block of wood, a sheet of metal, and a piece of black cloth are opaque substances.

The Color of Objects

Why is grass green, an apple red, and a daffodil yellow? The answer to this question depends on the nature of the object and the colors of the light striking the object.

If an object does not allow any light to pass through it (the object is opaque), the light falling on the object is either reflected or absorbed. If the light is absorbed, can it reach your eyes? Obviously not. Only the light that is reflected reaches you. So the color of an opaque object is the color it reflects.

Think for a moment of a red apple. A red apple reflects red and absorbs all the other colors. You see the red apple only by the light it reflects. What color do the green leaves and the stem of the apple reflect?

Now think about an object that is white. White is the presence of all the colors of the visible spectrum. So what is being reflected from a white object? You are right if you said all the colors are reflected. No color is absorbed. If, on the other hand, all the colors are absorbed, then no color is reflected back to you. The object appears black. Black is the absence of color. Have you ever noticed that many items of clothing, such as tennis clothes, that are worn in warm weather or in sunlight are made of white material? This is because white reflects the sunlight and thus the heat, keeping you cooler. Why are many cold-weather clothes made of dark-colored materials?

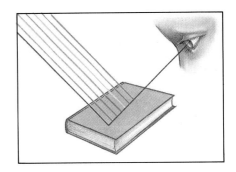

Figure 4–18 *The color of an object is the color of light reflected from the object to your eye.*

Figure 4–19 *In white light, the apple appears red because it reflects red light. If only green light shines on it, the apple appears black. Why?*

ACTIVITY

WRITING

Color Photography

Color film is sensitive to the various frequencies of light. It can record the colors of an object.

Using books and other reference materials in your library, find out how color film works. Write a report that describes how the various colors are recorded on the film.

Activity Bank

Spinning Wheel, p.147

Figure 4–20 *This plus sign looks quite colorful, but on a television screen it is actually white. The colors of images on a television screen are produced by combinations of tiny dots of the primary colors of light. The primary colors of light combine to form white light.*

If an object allows light to pass through it, the color that is transmitted is the color that reaches your eyes. The other colors are absorbed. So the color of a transparent object is the color of the light it transmits. Red glass absorbs all colors but red, which it transmits. Green glass transmits only green light. Ordinary window glass transmits all colors and is said to be colorless.

Adding Primary Colors of Light

Three colors can be mixed to produce light of any other color. These colors are called the primary colors. The primary colors of light are red, blue, and green.

Adding the proper amounts of these three colors produces any color, including white. For example, red and green light add together to produce yellow light. Blue and red add together to make magenta.

The color picture on a television screen or a computer screen is produced by adding the primary colors together. A screen contains groups of red, blue, and green dots. The correct mixture of these dots makes the pictures that you see.

Subtracting Colors of Pigments

You have just learned that all the colors of light combine to form white. But have you ever mixed all the colors in a paint set? Was the mixture white? Definitely not.

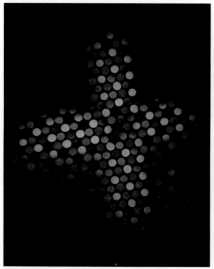

Paints are not sources of light. Paints are pigments. And the blending of pigments is different from the way your eyes blend the various colors of light. The primary pigments are yellow, cyan, and magenta. When the three primary pigments are mixed in equal amounts, all colors are absorbed and the result is black. Each different-colored pigment absorbs at least one color out of the visible spectrum of light that hits it. By mixing more and more pigments together, more of the visible spectrum is absorbed, or subtracted, from what you see. A movie film is made in this way.

Pigments have a certain color because they can absorb only certain wavelengths of the visible spectrum. All the rest are reflected. Does this sound similar to the explanation of why an apple is red? There is a good reason for the similarity. All objects contain pigments. The color of an object is a result of the pigments it contains.

Figure 4-21 *The primary pigments can be combined to form all the other colors. The pure colors can be lightened by adding white. Mixing together the three primary pigments produces black.*

Polarized Light and Filters

You can probably recall several instances in which you found yourself squinting because of the glare of the sun. But have you ever put on a pair of polarizing sunglasses to make the problem disappear? Sunglasses that reduce glare use polarizing filters.

You have learned that the waves of light from a normal light source vibrate in different directions at the same time. A polarizing filter is made up of a large number of parallel slits. When light passes through a polarizing filter, only those waves vibrating in the same direction as the slits can pass through. The light that passes through the filter is

Activity Bank

Color Crazy, p.149

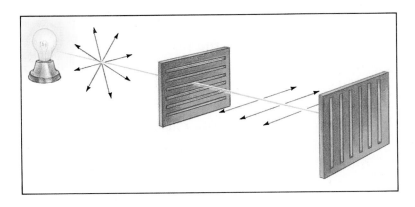

Figure 4-22 *Light waves that vibrate in all directions can be passed through a special filter to produce light waves that vibrate along a single plane. This light is called polarized light. What is a common use of polarizing filters?*

Figure 4–23 *Because of glare, all you can see are reflections on the surface of the water. When viewed through a polarizing filter, however, you can see the more detailed structure of the branch below the water.*

ACTIVITY

Colors and Filters

1. Obtain six color filters: red, green, blue, cyan, yellow, and magenta.

2. Look at a source of white light through each filter. Notice the colors you see.

3. Put the green filter behind the red filter. Look at the light through both filters.

4. Repeat step 3 for all combinations of filters.

■ How do the filters and combinations of filters affect the colors you see? Explain how color filters can be used to make slides.

■ How do you think tinted or colored eyeglasses or sunglasses affect what you see?

said to be **polarized light.** The rest of the light is either reflected or absorbed.

The glare of sunlight that makes you squint is mostly horizontal. Polarized sunglasses are vertically polarized. So the glare is reduced without affecting the rest of the light that enters your eyes.

Different types of filters are used to block certain colors of light. A color filter is a material that will absorb some wavelengths of light while allowing others to pass through. A common use of color filters is in slides. A color slide contains three filters. When white light from the projector is sent through the slide, each filter absorbs some colors and allows the other wavelengths to pass through. The combination of filters allows each image to appear in its natural color.

4–4 Section Review

1. Compare transparent, translucent, and opaque substances.
2. What are the primary colors of light? Of pigments?
3. What happens when the primary colors of light are mixed in equal amounts? When the primary pigments are mixed in equal amounts?
4. What is polarized light?

Critical Thinking—*Applying Concepts*

5. Why would a purple-people eater appear black under yellow light?

PROBLEM ??? Solving

Round and Round It Goes

You have been given a very strange-looking but intriguing device and only one instruction: Place it in sunlight. At first glance it appears to be useless. Upon closer inspection, you notice that it consists of four squares, and each square is silver on one side and black on the other. You're still not impressed, however.

So you place the contraption in front of a window. Suddenly it starts spinning wildly. The brighter the sunlight, the faster the device spins.

Making Inferences

1. Explain why the act of placing this device in sunlight makes it move.

2. What practical application might this device have?

4–5 How You See

You have learned what light is, how it is produced, how it reflects and refracts, and how colors appear. But how do you see light? **You see light through a series of steps that involve the various parts of the eye and the brain.**

Light enters the eye through an opening called the **pupil.** The pupil is the black circle in the center of your eye. It is black because no light is reflected from it. (Remember, black is the absence of color.) The colored area surrounding the pupil, called the **iris,** controls the amount of light that enters the pupil. When the light is dim, the pupil is opened wide. When light is bright, the pupil is partially closed, or small.

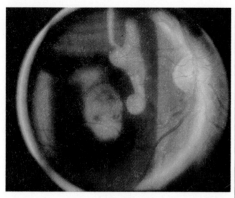

Figure 4–24 *The eye is the organ of sight. This photograph is an image as it is formed on the retina. What is true about the position of the image on the retina?*

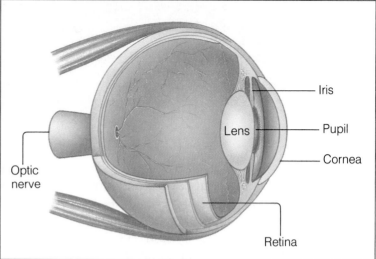

ACTIVITY

Color "Magic"

1. Stare at a star in the lower right of the flag for at least 60 seconds. Do not move your eyes during that time.

2. Quickly look at a sheet of blank white paper.

What image appears?

Explain why you see what you do.

Hint: When you stare too long at a color, the cones in your eyes get tired. You see the color that mixes with the original color to form white.

The light that enters the eye is refracted and focused on the curved rear surface of the eye, called the **retina.** The image that falls on the retina is upside down and smaller than the object. Most of the refraction is done by the protective outer covering of the eye, called the **cornea.** The lens of the eye makes adjustments for focusing on objects at different distances. Muscles attached to the lens control its shape. To see distant objects, the muscles relax, leaving the lens thin. To see nearby objects, the muscles contract, producing a thick lens.

The image that falls on the retina is then transferred to and interpreted by the nervous system. The retina is made of light-sensitive nerves that transfer the image to the brain. Some of the nerve cells in the retina are called **rods.** Rods are sensitive to light and dark. Other nerve cells called **cones** are responsible for seeing colors. Each cone is sensitive to a particular color.

Lenses and Vision

The lens of your eye is a convex lens. It is not a hard, rigid lens but rather a soft, flexible one. So it can easily change shape to allow you to see clear images of objects both near and distant.

NEARSIGHTEDNESS Ideally, the image formed by the lens should fall directly onto the retina. If the eyeball is too long, the image forms in front of the retina. This condition is called **nearsightedness.** A

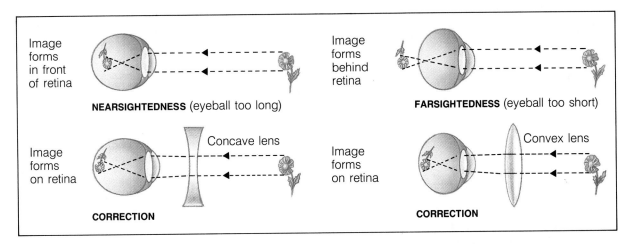

nearsighted person has difficulty seeing objects at a distance but has no trouble seeing nearby objects.

The lens of a nearsighted person is too convex. The rays of light converge at a point in front of the retina. A correcting lens would have to make the light rays diverge before they enter the eye. So a concave lens is used to correct nearsightedness.

FARSIGHTEDNESS If the eyeball is too short, the image is focused behind the retina. This condition is called **farsightedness.** A farsighted person can see distant objects clearly but has difficulty seeing nearby objects. These objects appear blurred.

The lens of a farsighted person is not convex enough. The rays of light converge at a point behind the retina. A correcting lens would have to make the rays converge before they enter the eye. So a convex lens is used to correct farsightedness.

4-5 Section Review

1. Describe the parts of the eye and their role in vision.
2. Where in the eye is the image formed?
3. What is nearsightedness, and how is it corrected? What is farsightedness, and how is it corrected?

Critical Thinking—*Making Inferences*
4. What causes colorblindness?

It's All an Illusion

Look at the two figures of circles. The center circle in the figure on the right is larger than the center circle on the left. Right? Wrong! Both circles are the same size. Sometimes your eyes play tricks on you. The way that you perceive an image to be is different from its true measurement, color, depth, or movement. This is called an *optical illusion.*

Some optical illusions are so convincing, you cannot believe it until you actually measure the objects or separate the colors. Look at the various figures shown below. Can you determine what is real and what is illusion?

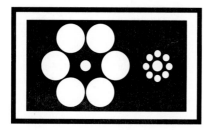

In the case of the circles, your brain compares the size of the center circle with those around it. Because the outer circles in the figure on the right are smaller, the center circle appears larger. A similar optical illusion occurs when you look at the moon or the sun. If it is close to the horizon, your brain compares it with houses and trees. It then appears much larger.

Now look at the spiral. Ooops, caught you again. The figure actually shows a series of circles. The background causes your eyes to perceive the circles as spiraling into the center.

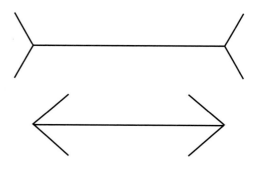

4-6 Optical Instruments

Reflection and refraction of light can be useful. Certain devices, known as optical instruments, produce images through reflection and refraction of light. Optical instruments use arrangements of mirrors and lenses to produce images.

Cameras

Like the eye, a camera works by allowing light to enter through a lens or a series of lenses. The major difference is that a camera permanently records an image on film. The film contains chemicals that are sensitive to light and that undergo change when light strikes them. Where does the image produced by the eye fall?

A camera contains a convex lens. When the shutter is opened, light from objects that are being photographed is focused by the lens as an image on the film. The image formed by a camera is real and upside down. It is also smaller than the actual object. The size of the image depends on the focal length of the lens.

It is important that not too much or too little light reaches the film. The opening in the camera through which light enters is called the aperture. The size of the aperture controls the amount of light that passes through the lens and reaches the film. What part of the eye plays the role of the aperture?

Guide for Reading

Focus on this question as you read.

▶ *What is the role of mirrors and lenses in cameras, telescopes, and microscopes?*

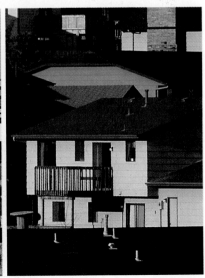

Figure 4-26 *These three photographs were taken with lenses of different focal lengths. The size of the image depends on the focal length of the lens. Which photograph was taken with a camera lens of short focal length? Of long focal length?*

ACTIVITY

DISCOVERING

Scope Out the Situation

1. Obtain two convex lenses, one of which is more curved than the other.

2. Hold the more curved lens near your eye. Position the other lens an arm's length away.

3. Select an object and move the farther lens toward or away from the object. Keep moving the lens until you see the object clearly through the combination of lenses.

Describe the image. Is it right side up or upside down? Is it larger or smaller than the object?

■ What kind of instrument have you made?

Telescopes

Telescopes are used to view objects that are very far away. Astronomers use telescopes to collect light from space. Because the light from many events that occur in space spreads out so much by the time it reaches Earth, the events are impossible to see with the unaided eye. The amount of light that can be gathered by the eye is limited by the eye's small size. By using large lenses and mirrors, telescopes can gather a great deal of light and capture images that otherwise could not be seen.

One type of telescope is called a refracting telescope. A refracting telescope consists of two convex lenses at opposite ends of a long tube. The lens located at the end of the telescope closest to the object is called the objective. This lens gathers light and focuses it to form an image. The other lens magnifies the image so that it can be observed by the eye. Sometimes the second lens is replaced with a camera so that the image can be recorded on film.

The larger the lens used in a refracting telescope, the greater the light-gathering power. But the construction of very large lenses is difficult. For this

Figure 4–27 *A refracting telescope uses a series of lenses to form an image of a distant object. The telescope at Yerkes Observatory in Wisconsin is the world's largest refracting telescope.*

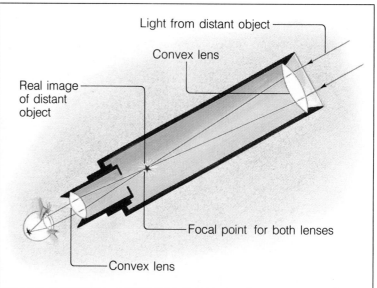

Light from distant object

Convex lens

Real image of distant object

Focal point for both lenses

Convex lens

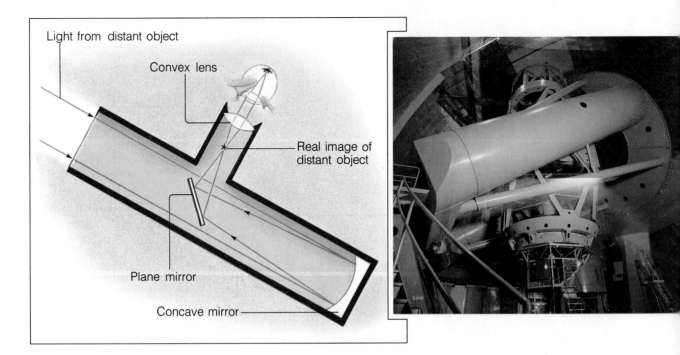

Light from distant object

Convex lens

Real image of distant object

Plane mirror

Concave mirror

Figure 4-28 *Light rays from a distant object are reflected from one mirror to another to produce an image in a reflecting telescope. The Hale Telescope at Mount Palomar Observatory in California is one of the world's largest reflecting telescopes.*

reason, telescopes were developed that use a large mirror instead of an objective lens. Large mirrors are easier to construct and support. A telescope that uses mirrors to gather light is called a reflecting telescope. The 200-inch Hale telescope at Mount Palomar in California is one of the world's largest reflecting telescopes.

Microscopes

When you want to get a better look at the detail of a nearby object, you bring it closer to your eye. But a person with normal eyesight cannot see an object clearly if it is closer than 25 centimeters from the eye. In addition, sometimes the object is just too small to see with the unaided eye. A microscope is used to aid the eye in magnifying nearby objects.

A microscope is similar to a refracting telescope, except that its goals are different. A microscope uses two convex lenses to magnify extremely small objects. One lens of the microscope, the objective, is placed at the end of the body tube close to the

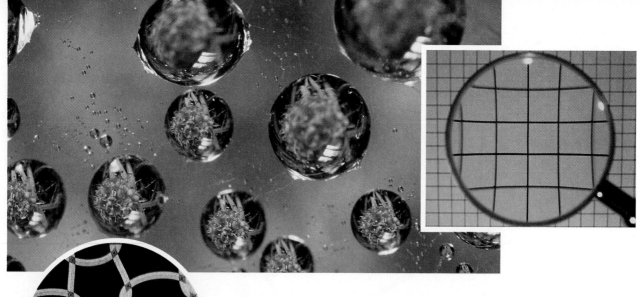

Figure 4-29 *These water droplets, captured by the web of a spider, show a clear image of a nearby flower. Droplets of water can act as convex lenses. A convex lens can also be used as a magnifying glass by placing the object within the focal length of the lens. A microscope uses two convex lenses to magnify images. This microscopic image shows the actual structure of a nylon stocking.*

object under inspection. The other lens, the eyepiece, is used to magnify the image formed by the objective. The total magnification produced by the microscope is the product of the magnification of the two lenses.

4-6 Section Review

1. Name three optical instruments. Explain what each does.
2. How are the parts of a camera similar to the parts of the eye?
3. Compare a refracting and a reflecting telescope.
4. How is a microscope similar to a refracting telescope? How is it different?

Critical Thinking—*Drawing Conclusions*
5. In some microscopes, oil rather than air is placed in the space between the two lenses. What effect might the oil have?

Guide for Reading

Focus on these questions as you read.

▶ What are the characteristics of laser light?

▶ How is laser light used in technology?

4-7 Lasers

A **laser** is a device that produces an intense beam of light of one color. The light given off by a laser is different from the light given off by an ordinary luminous object, such as the sun or a flashlight. Look at the light given off by the flashlight in Figure 4-30.

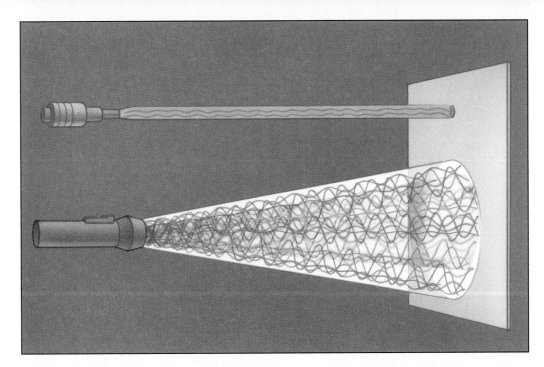

The light is a mixture of all the wavelengths of the visible spectrum. The light from the laser, however, is of only one wavelength. In the light given off by the flashlight, there is a mix of crests and troughs. The waves interfere, sometimes adding together and sometimes cancelling out. Eventually the light spreads out, decreasing its power. At a good distance from the flashlight, no light will be seen by an observer. In the laser, however, all the waves travel in step. The crests all travel next to one another and the troughs all travel next to one another. Light of the same wavelength that travels in step is said to be coherent light. **A laser is a device that produces coherent light.**

How Are Lasers Made?

In Chapter 3 you learned that electromagnetic waves are produced by moving charge. In the case of visible light, an electron in an atom absorbs a photon of energy and moves to a new position. The atom is said to be excited. When the electron releases the photon, it returns to its original position. But what happens if a photon hits an atom that is already in an excited state? The photon will not be absorbed by the excited atom. Instead, the photon will cause the excited atom to lose the energy it has already absorbed. The energy will be released in the form of a photon. Thus two identical photons that

Figure 4–30 *White light consists of a combination of wavelengths all traveling randomly. Laser light, however, consists of light of only one wavelength. All the crests and troughs of laser light travel together.*

Light Up Your Life

Your life would not be the same without the effects of light and its uses. Look around you and take note of the importance of light in your daily life. In addition to lighting up everything for you to see, consider the colors you notice and the instruments you use—including eyeglasses, telescopes, binoculars, and so on. Organize a list of at least ten examples. Explain how each example is made possible by light.

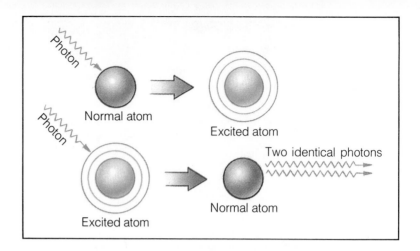

Figure 4–31 *When a photon hits a normal atom, its energy is absorbed and the atom becomes excited. When another identical photon hits the excited atom, the atom does not absorb the additional photon. Instead, it emits its extra energy so that two identical photons leave the atom, traveling in step with each other. What happens to the atom then?*

Figure 4–32 *Earlier scientists were often unable to test their scientific theories because the necessary technology did not exist. But today, researchers such as Dr. Raymond Chiao of the University of California at Berkeley have a wide variety of sophisticated devices at their disposal. This gives them the exciting ability to test and witness the predictions of their predecessors. Dr. Chiao has used lasers to experiment with light and to confirm theories proposed by Albert Einstein.*

are traveling in step with each other will leave the atom. If these two photons hit two more excited atoms, the process will be repeated, ending with four photons. Then four photons will hit four more excited atoms, and so on. This process is called stimulated emission. The word laser comes from *l*ight *a*mplification by *s*timulated *e*mission of *r*adiation.

A laser uses this process to produce a huge number of identical photons. A laser consists of a tube containing gases, liquids, or solids. The element used is chosen by the properties of the laser light it produces. Different elements produce different wavelengths of light. A mirror is placed at each end of the tube. The mirrors will cause any photons traveling down the tube to bounce back and forth between the ends. As they move back and forth, the photons will strike more and more atoms and keep producing millions of identical photons. One of the mirrors is a partial mirror. Instead of reflecting all the light that strikes it, it lets a tiny bit go through. As the beam becomes strong enough, a small percentage of the coherent light escapes through this mirror. This light becomes the laser beam.

Although there are many kinds of lasers, they all have certain features in common. They all consist of a narrow rod with mirrored ends containing a material that can produce identical photons. One of the earliest types of lasers was the solid ruby laser. A ruby laser consists of a solid rod made of ruby crystals. The most common type of laser is the helium-neon laser.

A laser also needs a source of energy that can put as many atoms as possible in an excited state. If the atoms are not excited, they will absorb the photons that hit them rather than emit more photons. Many lasers use electricity, but others use intense flashes of light or chemical reactions.

Uses of Lasers

Lasers have uses in medicine, manufacturing, communication, surveying, entertainment, and even measuring the distance to the moon. Lasers are used in audio and videodiscs, computers, and printers. In the future, lasers may be used to produce an almost limitless supply of energy for nuclear fusion.

MEASUREMENTS Laser beams can be made so strong and can be focused so well that they can be bounced off the moon. By carefully timing how long it takes for a laser beam sent from the Earth to reflect off a mirror on the moon's surface (put there by Apollo missions) and return back to Earth, the distance to the moon has been calculated within a few centimeters.

Because laser light travels in such a straight line, a focused laser beam makes an exceptionally accurate ruler. Lasers were used to lay out the tunnels of the Bay Area Rapid Transit (BART) system in California.

MEDICINE Sometimes the retina of the eye becomes detached from the inner surface of the eye and produces a blind spot. To treat this condition, surgeons use laser beams to focus light at the tip of the detached retina and thereby weld it back without cutting into the eye.

Surgeons also use lasers as knives to cut through human skin and tissue. Lasers are replacing scalpels in certain kinds of surgery. Lasers have the special advantage of heating while they cut. The heat closes the blood vessels, thus reducing bleeding.

INDUSTRY Do you remember the compact discs you read about at the beginning of this chapter? Compact discs are produced and interpreted by laser beams. When a CD is made, sound is converted to electrical signals. The electrical signals are fed to the laser, which emits light as flashes that cause a pattern of flat areas and pits on the disk. The disk is then coated for protection.

When the CD is played, another laser beam shines on the disc as it spins. A small device records the reflection of the beam. However, the reflection

Figure 4-33 *You may have seen lasers used to produce entertaining light shows or to read bar codes at the local grocery store. But laser light is also commonly used in surgery. This argon laser beam is passing into the bone of the ear.*

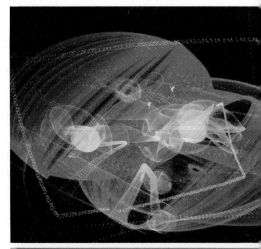

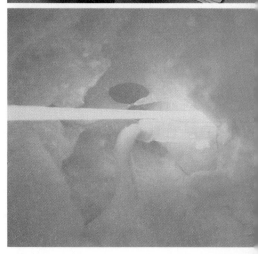

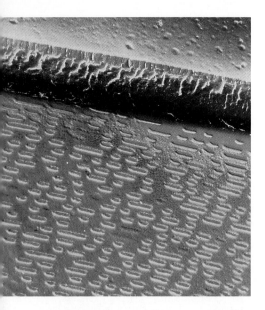

Figure 4–34 *The highly magnified surface of a compact disc shows the series of flats and pits that code the information it contains. The white portion at the top is its outer covering.*

Figure 4–35 *Fiber optic cables carry laser light for use in communication, medicine, and industry.*

is not constant, due to the flats and pits. The device reads the changes as a code that is converted to electrical signals. This time, the electrical signals are sent to speakers, which reproduce the original sound waves. Video discs carrying both sound and pictures are now being produced in a similar manner.

Most products you buy in the supermarket contain UPC codes (Universal Product Codes). These codes are read when a laser beam bounces off them. The information is fed into a computer, which automatically puts the price into the cash register. This process minimizes mistakes and helps keep track of inventory.

Fiber Optics

Imagine strands of glass, some of them ten times finer than a human hair, replacing copper wire in cables used to transmit telephone and television signals. The strands of glass are so pure that you can see clearly through a block of them 20 kilometers thick. The glass cables can carry more information at higher speeds than copper cable can—and they occupy only one tenth the space.

Pure imagination? Absolutely not! These strands of glass are called **optical fibers.** Optical fibers transmit signals as flashes of light. You learned in Chapter 3 that information is often carried on radio waves. But the amount of information that can be carried on an electromagnetic wave increases as the frequency increases. Since light has higher frequencies than radio waves, much more information can be carried on a light beam than on a radio wave. The study and use of optical fibers is called fiber optics.

TOTAL INTERNAL REFLECTION The principle that allows glass fibers to carry light involves both reflection and refraction. You know that when light crosses into a new medium in which its speed changes, it bends, or refracts. The greater the angle at which it enters the new medium (the angle of incidence), the more it is bent. If the angle of incidence is great enough, the light is bent so much that it is reflected rather than transmitted through the new medium. This type of reflection is known as **total internal reflection.** Very little light is lost when it is totally reflected.

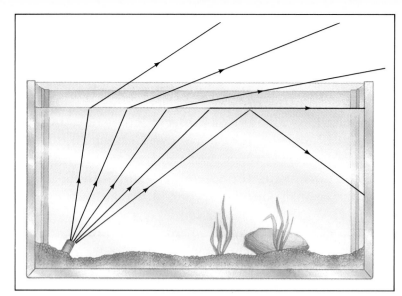

Figure 4–36 *Perhaps you have seen a light in the bottom of a fishtank or swimming pool. The angle at which the light is placed determines whether or not the light can leave through the top of the water. At any position greater than a certain angle, all light will be reflected back into the water. What is this principle called?*

Did you ever see a diamond sparkle brilliantly in the light? Diamonds would not sparkle if it were not for total internal reflection. In fact, raw diamonds are quite rough and dull. The angles cut into a diamond are precisely calculated so that all the light entering the diamond is reflected internally and emerges at the top. Diamond has a high index of refraction. This means that with proper cutting, all the light entering it can be reflected and directed to one point—the top. A poorly cut diamond will not sparkle. A diamond's brilliance is also caused by dispersion. The light that enters a diamond is broken up into the colors of the spectrum, making its reflections multicolored. In this case, what other object does a diamond act like?

You may be wondering what total internal reflection has to do with transmitting information through glass fibers. Well, it has everything to do with it! An optical fiber is like a "light pipe." If a beam of light is directed into one end of an optical fiber at the proper angle, it will always strike the walls at angles great enough to cause total internal reflection. Thus a ray of light is reflected so that it zigzags its way through the length of the fiber. Light entering one end of a fiber is internally reflected many times—as many as 15,000 times per meter—without being lost through the walls of the fiber. So signals do not fade away as easily as they do in copper wire.

When you talk on the telephone, your voice is converted into an electrical signal that goes to a tiny laser, no larger than a grain of salt. The laser transmits your voice as a series of flashes of light that

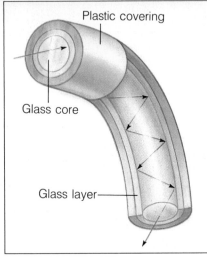

Plastic covering

Glass core

Glass layer

Figure 4–37 *By a continuous series of total internal reflections, laser light can be transported through an optical fiber. A tiny laser at the end of a fiber can shine light through a fiber, around a spool, and all the way to the end of the fiber.*

travel through the glass fiber. When you speak, you leave spaces between words and syllables. In the fiber, these spaces are filled with someone else's conversation. In this way, an optical fiber can carry a tremendous amount of information at one time. Optical fibers now in use can carry about 20,000 telephone calls at once! Just think of that the next time you are using the telephone. In addition, sound, pictures, and computer information can all be carried in the same cable.

The applications of fiber optics are not limited to communication. Advances in the field have made it possible for doctors to see inside a patient's body without having to perform surgery. Surgeons can snake a thin optical fiber through a portion of the body to a specific region such as the heart or the intestines. In this way, particular organs can be studied without the risks of surgery.

Holography

When you take a photograph of an object or scene, does the photograph turn out to be quite the same as you remember the scene to be? Probably not. The major difference between what you see and what a photograph records is dimension. You see objects in three dimensions. The reason for this is that light waves are reflected from every point on the object. At certain points, the reflected light waves overlap and interfere with one another. The interference pattern of the reflected waves gives the object its depth. A camera cannot capture that information. So photographs are only two dimensional.

There is a technology, however, that uses laser beams to form three-dimensional images. The technology is called **holography,** and the three-dimensional image is called a hologram. The word hologram comes from the Greek words *holos* and *gramma,* meaning whole message. Because holography captures all the information coming from an object, it provides depth as well.

Instead of recording an image, holography uses a laser beam to record the entire interference pattern created by the light reflected from an object—just as your eyes do. A laser beam is split into two parts. One part is sent directly to the film, and the other is sent to the object. The object then reflects this part of the beam to the film. At the film, the two halves of the beam interfere with each other, producing the same interference pattern that is recorded by your eyes.

A hologram is a record of the interference pattern. When you look at the piece of film that makes up a hologram, you do not see the image. In fact, a hologram looks more like a transparent piece of plastic with smudges on it. But when you shine the same type of laser beam that made the hologram through it, you see the object in three dimensions. What you are doing when you shine a laser beam through a hologram is making a set of light waves exactly like those originally reflected from the object. Just as with a real object, the image changes when you look at it from different angles.

Holography is currently being used in many interesting ways. Holograms are used on credit cards and clothing labels as security devices because the pictures are almost impossible to forge. Holograms are also used to test products for structural flaws. For example, technicians make holograms of new tires before and after subjecting them to stress. Putting one image on top of the other shows the flaws. Inside the human body, holograms can be used to give three-dimensional views of organs. And because holograms can store a tremendous amount of data in a limited space, they may someday be used to store books and other reference materials. Eventually, holography may even bring three-dimensional television pictures and artwork into your home.

Figure 4-38 *Holograms capture three-dimensional images on film. Notice the depth of the fruit hologram. By recording the interference pattern of light reflected off an object, such as a statue, a hologram can be made.*

4-7 Section Review

1. What is a laser? How is laser light made? How are lasers used?
2. Distinguish between white light and laser light.
3. What is total internal reflection? How is it related to fiber optics?
4. What are some applications of fiber optics?
5. What is a hologram? What are some uses for holography?

Critical Thinking—*Applying Concepts*
6. Why is it dangerous to look into a laser beam?

Laboratory Investigation

Convex Lenses

Problem

What kinds of images are formed by convex lenses?

Materials *(per group)*

convex lens
lens holder
light bulb and socket
blank sheet of paper
meterstick

Procedure ⚗ 🔬

1. Place the convex lens in the lens holder and position them at least 2 meters in front of the lighted bulb.

2. Position the paper behind the lens so a clear image of the bulb can be seen on the paper. The sheet of paper must be positioned vertically. Record the position and relative size of the image.

3. Measure the distance from the lens to the paper. This is the focal length of the lens. Record the distance in centimeters.

4. Turn off the light bulb.

5. Move the bulb to a position that is greater than twice the focal length of the lens. Turn on the light bulb. Record the position and relative size of the image.

6. **CAUTION:** *Turn off the light bulb each time you move it.* Move the bulb to a position that is exactly twice the focal length. Record the position and relative size of the image.

7. Move the bulb to a position equal to the focal length. Record the position and relative size of the image.

8. Move the bulb to a position between the lens and the paper. This position is less than one focal length. Record the position and relative size of the image.

Observations

Describe the image formed in each step of the procedure for which you have made observations.

Analysis and Conclusions

1. Is the image formed by a convex lens always right side up? If not, under what conditions is the image upside down?

2. What happens to the size of the image as the bulb moves closer to the focal length? To the position of the image?

3. What happens to the size of the image as the bulb moves to less than the focal length? To the position of the image?

4. **On Your Own** Design and complete an experiment to study the images formed by concave lenses.

Study Guide

Summarizing Key Concepts

4–1 Ray Model of Light

▲ According to the ray model, light travels in straight-line paths called rays.

4–2 Reflection of Light

▲ The bouncing back of light when it hits a surface is called reflection.

▲ A plane mirror has a flat surface. A concave mirror curves inward. A convex mirror curves outward.

4–3 Refraction of Light

▲ The bending of light due to a change in speed is called refraction.

▲ A convex lens converges light rays. A concave lens diverges light rays.

4–4 Color

▲ The color of an opaque object is the color of light it reflects.

▲ The color of a transparent object is the color of light it transmits.

4–5 How You See

▲ Light enters the eye through the pupil. The amount of light that enters is controlled by the iris. The cornea and the lens refract light to focus it on the retina. The brain interprets the image.

4–6 Optical Instruments

▲ The image formed by a camera is real, upside down, and smaller than the object.

▲ A refracting telescope uses lenses to focus and magnify light. A reflecting telescope uses mirrors to gather light.

▲ A microscope uses lenses to magnify extremely small objects.

4–7 Lasers

▲ Light from a laser has one wavelength and forms an intense, concentrated beam.

▲ Optical fibers are long, thin, flexible fibers of glass or plastic that transmit light.

▲ A hologram is a three-dimensional picture made from laser light.

Reviewing Key Terms

Define each term in a complete sentence.

4–2 Reflection of Light
regular reflection
diffuse reflection
plane mirror
concave mirror
focal point
convex mirror

4–3 Refraction of Light
index of refraction
lens
convex lens
concave lens

4–4 Color
transparent
translucent
opaque
polarized light

4–5 How You See
pupil
iris
retina
cornea
rod
cone

nearsightedness
farsightedness

4–7 Lasers
laser
optical fiber
total internal reflection
holography

Chapter Review

Content Review

Multiple Choice

Choose the letter of the answer that best completes each statement.

1. Shadows can be explained by the fact that light travels as
   a. waves.
   b. curves.
   c. photons.
   d. rays.
2. The scattering of light off an irregular surface is called
   a. diffuse reflection.
   b. refraction.
   c. regular reflection.
   d. total internal reflection.
3. The point in front of a mirror where reflected rays meet is the
   a. aperture.
   b. focal length.
   c. focal point.
   d. index of refraction.
4. Crests and troughs of each wave are lined up with crests and troughs of all the other waves in
   a. incandescent light.
   b. fluorescent light.
   c. neon light.
   d. laser light.

5. A substance that does not transmit light is
   a. translucent.
   b. opaque.
   c. transparent.
   d. polarized.
6. The colored part of the eye is the
   a. iris.
   b. pupil.
   c. retina.
   d. cornea.
7. A nerve cell that responds to light and dark is a(an)
   a. cone.
   b. iris.
   c. rod.
   d. pupil.
8. Farsightedness is corrected by a
   a. convex mirror.
   b. convex lens.
   c. concave mirror.
   d. concave lens.
9. The process by which white light is separated into colors is
   a. reflection.
   b. absorption.
   c. transmission.
   d. dispersion.

True or False

If the statement is true, write "true." If it is false, change the underlined word or words to make the statement true.

1. An image that only seems to be where it is seen is a <u>real</u> image.
2. <u>Convex</u> mirrors are placed behind car headlights to focus the beam of light.
3. A mirage is an example of the <u>reflection</u> of light by the Earth's atmosphere.
4. A lens that is thicker at the ends than in the middle is a <u>concave</u> lens.
5. Frosted glass is an example of a <u>transparent</u> substance.
6. A telescope that uses mirrors to gather light is a <u>reflecting</u> telescope.
7. A <u>laser</u> is a device that produces coherent light.

Concept Mapping

Complete the following concept map for Section 4–2. Refer to pages R6–R7 to construct a concept map for the entire chapter.

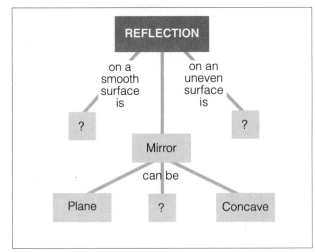

Concept Mastery

Discuss each of the following in a brief paragraph.

1. What is the ray model of light?
2. Compare a regular and a diffuse reflection.
3. Compare a real image and a virtual image. When you go to the movies, do you see a real or a virtual image on the screen?
4. When light passes from air into glass, which characteristic does not change: wavelength, frequency, or speed?
5. Explain why mirages form.
6. Explain why a spoon looks bent when you place it in a glass of water.
7. Describe the role of each part of the eye in sight.
8. How do polarized sunglasses help you to see in sunlight without squinting?
9. Why do you see a rose as red?
10. Describe two vision problems. How is each corrected?
11. What are two advantages of using large reflecting mirrors in astronomical telescopes?
12. What is the purpose of mirrors in a laser? Why is one less than 100 percent reflecting?
13. Explain how an optical fiber transmits light.
14. What is a hologram? How is it formed?

Critical Thinking and Problem Solving

Use the skills you have developed in this chapter to answer each of the following.

1. **Applying concepts** Why must the moon have a rough surface rather than a smooth, mirrorlike surface?

2. **Making calculations** The speed of light in a glass block is 150,000,000 m/sec. What is the index of refraction of the material?
3. **Applying concepts** A prism separates white light into the colors of the spectrum. What would happen if a second prism were placed in the path of the separated colors? Use a diagram in your answer.
4. **Making diagrams** Explain why you still see the sun setting just after it has truly set. Use a diagram in your answer.
5. **Identifying relationships** Why are roadways usually made of materials that are dark in color?
6. **Making inferences** Cones stop working in dim light. Explain why color seems to disappear at night.
7. **Making comparisons** Compare the structure and operation of a camera to that of the eye.
8. **Applying definitions** The lens at the end of a microscope tube has a magnification of $10\times$. The eyepiece lens has a magnification of $6\times$. What is the total magnification achieved by the microscope?
9. **Using the writing process** The technology is available to place a series of mirrors in orbit. These mirrors would reflect light back to Earth to illuminate major urban areas at night. Write a letter expressing your opinions about such a project. What do you think are some of the problems that would result from such a project? What are some of the benefits?

GAZETTE

ALLAN HOUSER:

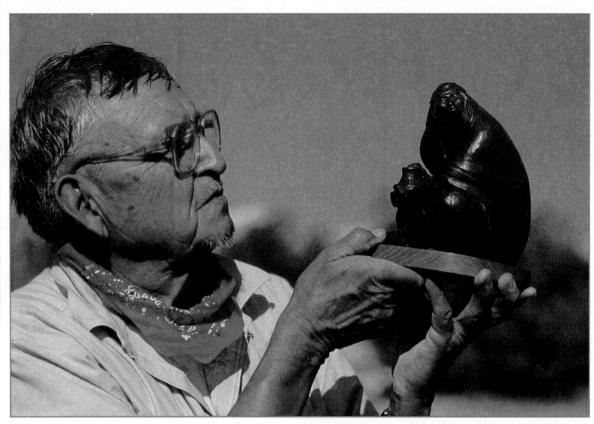

AN ARTIST'S ADVENTURE

Are you surprised to find a story about an artist in Adventures in Science? Perhaps you do not realize that a painter must be something of a scientist in order to use colors and light creatively. Sculptors also need to know scientific principles of force and balance to create a statue. One artist who has been both a painter and a sculptor is the Native American artist Allan Houser.

Allan Houser was born in Oklahoma to Apache Indian parents with the surname Haozous. His Apache background is evident in his appearance: keen glance, wide cheek-

bones, and sharply angled brows. His family was descended from a line of chiefs and leaders. In fact, his father, Sam Haozous, was one of Geronimo's warriors. Houser is what is known as a first-generation transitional Indian. This means that he has successfully built his life within the majority culture while still preserving a distinct sense of his Native American heritage.

Houser's parents lived and worked on a government-grant farm. For American Indians to survive as farmers was a continual challenge, and life was difficult. Although Allan began high school at Chilocco Indian School in northern Oklahoma, he had to drop out after the first year in order to work on the family farm. He never did return to high school. But after five years of working on the farm, he was able to begin art studies at the Painting Studio of Santa Fe Indian School.

Houser's career as a painter developed rapidly. After just two years of art school, he gained international recognition by having his paintings on display at the 1936 New York World's Fair. Three years later, he was commissioned to paint murals in the Department of the Interior building in Washington, DC. That same year, his paintings were displayed at the National Gallery of Art in Washington and at the Art Institute of Chicago.

Houser began sculpting in 1941. By this time, however, he was faced with financial problems. In order to support his family, he took a job as a construction worker and pipe-fitter's assistant. He continued painting and sculpting at night. After long years of struggle, in 1949 he received a Guggenheim Fellowship. With the fellowship, he once again had the freedom to devote himself to his work as an artist.

Houser draws much of the inspiration for his work from nature and the land. Granite cliffs, crumbling sandstone walls, rolling foothills, and valleys scooped by wind and rain all find their way into his painting and sculpture.

His love of nature keeps Houser in New Mexico. He often speaks of "doing research" there in much the same way that a scientist does. And, indeed, he is something of a scientist as he studies interesting rocks, fossils, and bones—especially skulls. Houser explains, "There are many very sculptural pieces in a skull. You can pull them apart and find unusual things, like the jaw bone [which is] a very strong form."

Since the beginning of his career, Houser's work has been drawn entirely from Native American themes. Yet, like all great art, his work transcends race and language. Houser's work is a link between modern life and the spirit of primitive culture. As a result, he has been able to do a great deal to advance the status of work by Indian artists.

Today, Houser is considered by many to be the greatest Native American artist. He has received many honors, but none that means more to him than the installation of one of his bronze sculptures in the United States Mission at the United Nations. Entitled *Offering of the Sacred Pipe*, the statue depicts a slender figure clothed in a feathered headdress and a cloak of animal skin. The figure holds aloft a sacred pipe as an offering to the Great Spirit. The strong upward lines of the figure symbolize humanity's search for lofty ideals, hope, and expanded potential. Former United Nations Ambassador Jeane Kirkpatrick described the statue as "A prayer for peace... a symbol of this country's identity and roots."

When he is not working as an artist, Houser enjoys playing music. So, as you can see, every aspect of Allan Houser embodies a wondrous spirit filled with vigor and life. His work and his life are composed of a beautiful union of color, light, sound, and nature.

▼ **Standing proudly outside the United Nations building, this bronze statue is a tribute to the strong spirit of Allan Houser and all that he believes in.**

NOISE POLLUTION: Is Your Hearing in Danger?

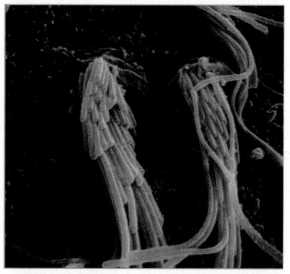

▲ **Once damaged, the delicate hair cells in the ear can no longer receive sounds and send messages to the brain. Thus hearing becomes impossible.**

Has the constant noise from a construction site, or even the traffic outside your home, ever annoyed you to the point where you became cranky and irritable? Or perhaps a baby's continual crying disturbed you in a restaurant or in an airplane? If so, you are not alone in your frustrations with unpleasant noises, and your complaints are not new. As far back as the first century BC, the emperor Julius Caesar passed a law forbidding chariots in Rome because of the distracting clatter made by their wheels as they rolled across the city's stone and dirt roadways.

Today, however, with the improvements and developments in technology, attempts to eliminate noise are not as easy. Nonetheless, since the days of Caesar many people and governments have taken actions and passed laws to prevent excessive noise. But does noise really deserve such attention? Is it really a hazard or are some people just not willing to tolerate noise?

According to sound and hearing specialists, noise does warrant serious concern. Sound travels as a wave that carries energy from one place to another. The louder the sound, the more energy it carries. Experiments show that exposure to loud or continuous noise can deliver enough energy to human cells to damage tissues, particularly those in the ear. For example, experts say that either a single dose of a loud noise or prolonged exposure to a strong sound can cause permanent hearing loss. The pressure of intense vibrations can cause the eardrum to tear or can damage the fragile hair cells responsible for detecting and relaying sound messages to the brain.

Exposure to excessive noise is taking its toll in the United States. Specialists estimate that 1 in 10 people (around 30 million) have experienced some degree of hearing loss from common sounds. Noise that threatens human health is more than a mere nuisance: It is a pollutant! The effects of noise on the environment are not unlike those of smog, insecticides, and automobile exhaust.

Such high numbers of people with hearing loss may surprise you. This is because when you consider dangerous noises, you probably think of factories, farms, and construction sites. Certainly the hazards from industrial noises associated with heavy equipment are genuine causes for concern. What you may not have considered as noise pollutants, however, are your home, your neighborhood, or the entertainment you enjoy. Constant exposure to the sounds of modern appliances, from alarm clocks and hair dryers to vacuums and stereos, and the sounds of recreational activities, from roaring crowds to the explosion of hunting guns, is just as dangerous.

Most household sounds would be harmless if they occurred one at a time, infrequently, or over short periods of time. But taken together, the collection of household sounds can add up to 14 or 16 hours of intense noise every day. Barking pets, gardening tools, and rumbling automobiles also contribute to your daily dose of sound.

The most common culprit of recreational noise pollution is loud music. You have probably read about rock stars who have lost their hearing due to prolonged exposure to loud music. But with today's personal stereos, rock concerts are no longer the only source of musical dangers. A recent study of high school students suggests that loud music is wreaking havoc with students' ears.

Noise pollution may be as harmful to your emotions as it is to your ears. Exposure to noise can cause people to feel tense and anxious, especially when the noise is unwanted. In this way, noise adds to other factors that cause stress and its accompanying symptoms—headaches, ulcers, and heart ailments. Scientists have even suggested that exposure to loud noise may be associated with high cholesterol levels.

▼ **The loud music emitted by a personal stereo not only endangers the person who carries the stereo, but also the unfortunate bystanders exposed to the noise through no fault of their own.**

▲ The protective coverings on his ears are not simply to make his job more comfortable. They are essential to his health. Without ear protection, the tremendous noise that exists on an aircraft carrier would destroy a person's hearing in a relatively short amount of time.

In addition, noise pollution can be responsible for disturbing sleep. Even noises that do not awaken a person can disrupt sleep, making the person tired and irritable in the morning. People who have this type of disrupted sleep are often less sympathetic to the problems of others, quick to anger, and extremely distracted. Some people even lose their creativity and their ability to solve problems. In fact, studies have shown that students who come from quiet home environments have better attention spans and powers of concentration than do those who come from loud environments.

What can be done to control noise pollution? It is hoped that laws will be passed and technological improvements will be invented that will stop noise at its sources. But you can help protect yourself and your family right now. You can correct some noises in your home. For example, insulated win-

dows and carpeted floors absorb many sounds before they reach you. Sealing cracks or holes in walls and roofs can also prevent sound from entering your home. And you can limit the intensity of the sound you hear when listening to music or watching television.

One surprising example of protection from noise pollution is occurring in England. Elephants in the zoo wear huge earmuffs to protect their ears from the noise of the nearby airport. Noise pollution must be taken seriously in order to prevent a time when people, too, have to wear earmuffs just as they wear shoes and socks.

▼ Although this baby elephant is rather adorable with its big earmuffs, it is sad to think that the world is becoming so dangerously noisy that animals can no longer roam around freely and safely.

SAILING TO THE STARS

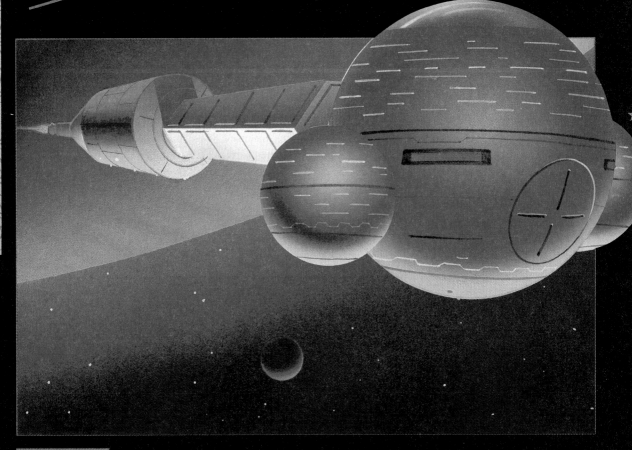

Tim gulped down his breakfast on Starship *Metropolis* and headed for the space school on Deck 2. He was late. In another 10 minutes, the first-period bell would ring and the space-technology test would begin. Today's test was on the history of starships. Tim was counting on his own experience with space technology to help him pass the test. After all, he had lived on a spaceship all his life.

Tim had been born 14 years before on Starship *Metropolis*. The ship was then in the fifteenth year of its 40-year voyage to Earth's nearest star, Alpha Centauri. His father and mother had been part of the original crew that had boarded the spaceship on Earth in the year 2079. But for Tim, Earth was almost an unknown planet. Certainly his parents still talked about Earth, and he was mildly curious. But for the most part, Earth was only the subject of a history course that he would study the following year.

Tim's world was the 2-kilometer-long Starship *Metropolis*. It was made up of 20 circular decks containing 2500 passengers, 20 shops, a hospital, a laser video theater, and 1100 living units. Most living areas were surrounded by constantly rotating cylinders that prevented weightlessness. The rotating cylinders also shielded Tim and the other people on the ship from the dangerous and penetrating radiation of space.

NUCLEAR POWERED ROCKETS

Tim trotted along the circular corridor of Deck 11. As he ran, he reviewed some of the facts about early space technology. He knew that the first spaceships had been based on engineering principles developed by Earth scientists more than a hundred years ago. That time was now known as the "great age of space engines." During this time, the first chemically powered rockets were produced. These rockets were able to escape the pull of Earth's gravity and reach the moon. Be-

fore the end of the twentieth century, however, most space scientists knew that no spaceship could travel fast enough to reach the distant stars without the use of nuclear fuel. Nuclear fuel is millions of times more powerful than any chemical fuel.

Deuterium, a heavy form of hydrogen, was the first nuclear fuel used in the spaceships of the early twenty-first century. The deuterium atoms were fused, or brought together, in nuclear reactors. This fusion process produced an almost unending flow of energy. Some of the early nuclear spaceships were capable of carrying several million tons of deuterium for long voyages to stars outside the Earth's solar system. Others carried less fuel but contained refining equipment. This equipment could get deuterium out of raw materials that the ship stopped to mine on other planets during the space voyage.

By the time Tim reached Deck 5, where the shops were arranged like a shopping mall, he was beginning to feel more confident about

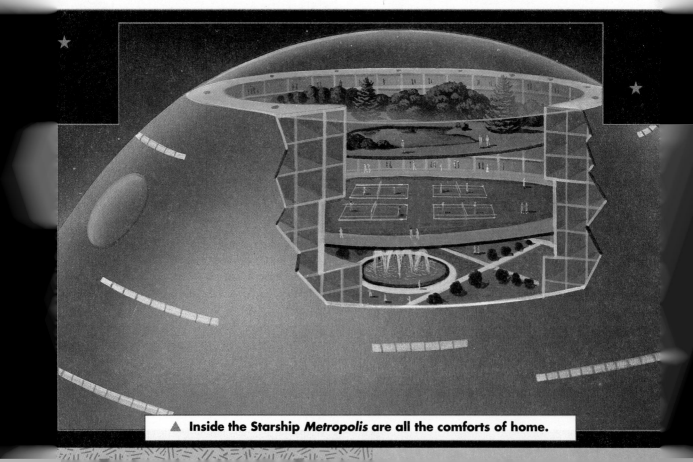

▲ Inside the Starship *Metropolis* are all the comforts of home.

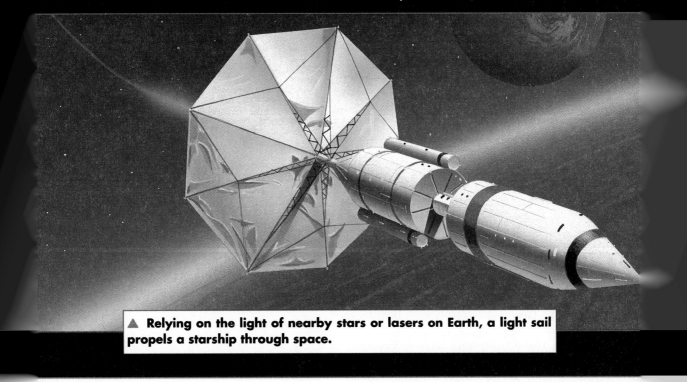

▲ Relying on the light of nearby stars or lasers on Earth, a light sail propels a starship through space.

the test. If enough of the test questions turned out to be about his own spaceship, he knew he would have no trouble passing.

THE LATEST IN STARSHIPS

Starship *Metropolis* represented one of the most amazing achievements in spaceship technology. It could go anywhere without ever running out of fuel. It would never run out of fuel because it was powered by hydrogen that it picked up from space. At this very moment, the ship was plunging through space at more than one tenth the speed of light. It was attracting hydrogen from space with a magnetic field shaped like an enormous funnel. The magnetic field was thousands of kilometers in diameter. When the ship finally approached the end of its journey, it would be slowed down by reversing its magnetic field to repel, rather than attract, hydrogen.

Between Decks 3 and 2, where the classroom was located, Tim came to a passage without gravity. He dropped into a suction chute and slid into the weightlessness of the passage. There he floated down the passage, using handles on the wall to move himself

along. While he glided, he felt a tinge of pride about living on Starship *Metropolis*. It certainly was the latest in spaceship technology. Then he remembered the new starship being designed on Earth, which he had learned about in class. The new ship would be propelled by a light sail made of a thin sheet of aluminum that was many kilometers wide. As this ship traveled through space, the sail would gather light energy from a laser on Earth, or the light from nearby stars. The light energy would be used to send the ship sailing through space.

Tim reached Deck 2 and cut across the park, which contained full-sized trees and an entire flower garden in bloom. For a moment he was struck by the fact that his entire world, including apartments, a school, and even a park, could be contained in one starship hurtling through space. What would it be like, he wondered, when the journey finally came to an end and he was able to leave the only world he had ever known?

Tim looked up through the plastic dome of the ship. A very bright yellow star shone in the blackness of space. Alpha Centauri was like a beacon guiding Tim and the Starship *Metropolis* to a new home.

For Further Reading

> If you have been intrigued by the concepts examined in this textbook, you may also be interested in the ways fellow thinkers—novelists, poets, essayists, as well as scientists—have imaginatively explored the same ideas.

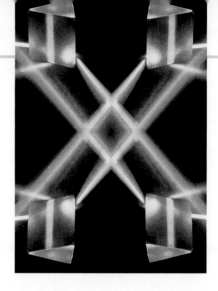

Chapter 1: Characteristics of Waves

Brandon, Fran. *Earthquake: The Day the Woods Went Crazy.* Nashville, TN: Winston-Derek.

Callahan, Steven. *Adrift: Seventy-Six Days Lost at Sea.* Boston: Houghton Mifflin.

Christopher, Matt. *Earthquake.* Boston: Little, Brown.

Forester, C. S. *Commodore Hornblower.* Boston: Little, Brown.

Olney, Ross. *A Young Sportsman's Guide to Surfing.* Nashville, TN: Thomas Nelson.

Stevenson, Robert Louis. *Treasure Island.* New York: Grosset & Dunlap.

Verne, Jules. *Twenty Thousand Leagues Under the Sea.* New York: New American Library.

Yolen, Jane. *Neptune Rising: Songs and Tales of the Undersea Folk.* New York: Philomel Books.

Chapter 2: Sound and Its Uses

Busnar, Gene. *The Superstars of Rock: Their Lives and Their Music.* Englewood Cliffs, NJ: Messner.

Christian, Mary B. *Singing Somebody Else's Song.* New York: Macmillan.

Danziger, Paula. *There's a Bat in Bunk Five.* New York: Dell.

De Veaux, Alexis. *Don't Explain (A Song of Billie Holiday).* New York: Harper & Row Junior Books.

Goffney, Timothy R. *Chuck Yeager: First Man to Fly Faster Than Sound.* Chicago: Children's Press.

Landis, J. D. *The Band Never Dances.* New York: Harper & Row Junior Books.

Neimark, Anne E. *A Deaf Child Listened: Thomas Gallaudet, Pioneer in American Education.* New York: William Morrow.

Salerno-Sonnenberg, Nadja. *Najda on My Way.* New York: Crown.

Chapter 3: Light and the Electromagnetic Spectrum

Benchley, Nathaniel. *The Electromagnetic Spectrum: Key to the Universe.* New York: Harper & Row Junior Books.

Branley, Franklyn. *Bright Candles.* New York: Harper & Row Junior Books.

Faraday, Michael. *Faraday's Chemical History of a Candle.* Chicago: Chicago Review.

Keller, Mollie. *Marie Curie.* New York: Franklin Watts.

Zindel, Paul. *The Effect of Gamma Rays on Man-in-the-Moon Marigolds.* New York: Bantam Books.

Chapter 4: Light and Its Uses

Butler, Beverly. *Light a Single Candle.* New York: Pocket Books.

Calvert, Patricia. *Yesterday's Daughter.* New York: Macmillan.

Collura, Mary-Ellen Lang. *Winners.* New York: Dial.

Kipling, Rudyard. *The Light That Failed.* New York: Airmont.

McDonough, Jerome. *Mirrors.* Schulenburg, TX: I. E. Clark.

O'Neill, Mary. *Hailstones and Halibut Bones.* New York: Doubleday.

Taylor, Theodore. *Walking Up a Rainbow.* New York: Dell.

Wells, H. G. *The Invisible Man.* New York: Bantam Books.

Woolgar, Jack. *Mystery in the Desert.* Jamaica, NY: Lantern.

Activity Bank

Welcome to the Activity Bank! This is an exciting and enjoyable part of your science textbook. By using the Activity Bank you will have the chance to make a variety of interesting and different observations about science. The best thing about the Activity Bank is that you and your classmates will become the detectives, and as with any investigation you will have to sort through information to find the truth. There will be many twists and turns along the way, some surprises and disappointments too. So always remember to keep an open mind, ask lots of questions, and have fun learning about science.

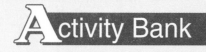

A wave originates when a vibrating source gives it energy causing it to vibrate as well. This may be easy to remember, but it can be difficult to understand. After all, it is often hard to picture something that you cannot see. In this activity, however, you will have an opportunity to create and indirectly observe the vibrations of sound waves.

What You Need

cardboard cylinder, such as from a container of oatmeal
balloon
rubber band
sugar or salt
wooden spoon
metal baking tray
scissors
graph paper

What You Do

1. Remove both ends from the cardboard cylinder.

2. Cut the balloon and stretch it tightly over one end of the cylinder. Secure it by stretching the rubber band around the balloon and the cylinder. It is important that the balloon be stretched tightly for this to work. You have made a simple drum.

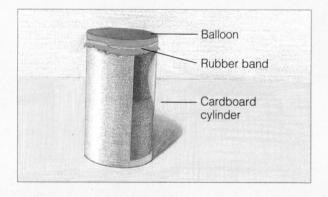

Balloon

Rubber band

Cardboard cylinder

3. Sprinkle sugar (or salt) on the drum head (stretched balloon). Tap the edge of the drum head several times with the wooden spoon. What happens to the sugar? Hit it harder and observe any further changes.

4. Spread the sugar out on the drum head again. Have a classmate hold the drum head level while you shout into the open end. What happens to the sugar this time? Use graph paper to draw your observations. Repeat this step, but this time have your classmate shout and you observe what happens. Again draw your observations.

5. Spread the sugar out over the drum head once more. At a distance of about 20 cm above the drum head, tap a metal baking tray with the wooden spoon. What happens to the sugar?

Observations and Conclusions

1. What do you observe when you tap on the drum head in step 3?

2. How can you explain what happens to the sugar or salt?

3. What gave energy to the drum head?

4. What happens when you shout into the cylinder in step 4? Is energy involved this time?

5. Compare the graph you made when you shouted into the cylinder with those of your classmates. Are they alike? Do they show any differences?

6. What do you observe when the baking tray is hit with the wooden spoon in step 5?

7. Explain how the metal tray being hit can affect the drum without even touching it.

SOUND AROUND

You know that a ball will bounce off a sidewalk and that light will reflect off a mirror. But will sound bounce off a hard surface as well? Because sound is a wave and waves are reflected, it should. But by doing this activity you can prove it for yourself!

Materials

two sturdy cardboard tubes without ends (poster or art tubes, for example)

small wind-up toy or ticking watch

large smooth piece of sturdy cardboard or plastic, about 40 cm × 40 cm

pieces of fabric and foam

Procedure

1. Have a classmate hold the cardboard or plastic upright on a large table or on the floor.

2. Place the tubes on the surface of the table or floor at an angle to the cardboard and to each other. Leave a gap of about 6 cm between the cardboard and the ends of the tubes.

3. If the tubes have caps, place a cap on the outer end of one tube. If the tubes do not have caps, ask a second classmate to cover the outer end of one tube with his/her hand or a book. Listen through the end of the uncovered tube. Cover your outer ear with your hand so that the only sounds you hear are coming through the tube. What do you hear?

4. Have your classmate uncover or take the cap off the other tube. Place the watch or toy just inside the end of the tube and put the cap back on the tube or cover it again. Listen in the other tube as you did before. What do you hear? Explain your observations.

5. Switch places with your classmate so that you cover the tube and your classmate listens. What does your classmate hear in each case?

Going Further

Repeat the activity several times, but each time cover the cardboard with a different material, such as a variety of fabrics and foam. Determine how these materials affect what you hear. Explain why.

■ Why is a carpeted room more quiet than an uncarpeted room?

■ Some automobiles are very noisy when you ride in them. How can automobile manufacturers make cars less noisy?

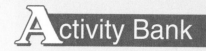
PRESTO CHANGO, IT'S GONE

Do you think that you can make an object disappear before your very eyes? You can—and in this activity you will find out how! The secret is understanding refraction—in this case the bending of light as it travels from one medium into another.

You will need an empty jar with a lid (preferably a short jar such as a peanut-butter jar), a postage stamp, and some water.

Procedure

1. Put the stamp on a tabletop.
2. Place the jar, open end up, over the stamp. Look at the stamp.
3. Fill the jar with water and put the lid on it.
4. Look at the stamp now.

What You Saw

1. What did you observe when you placed the jar over the stamp?
2. Did adding water to the jar affect how you saw the stamp?
3. Can you explain your observations?
4. How is this activity different from the one that appears at the end of the first paragraph on page R26? Draw diagrams in your explanation.

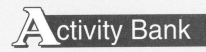
BELLS A' RINGING

Can you imagine shouting as loud as you can yet not making any sound? This is exactly what would happen if you tried to shout in outer space. And the reason is that in space there are no air molecules for sound to travel through. Although you cannot reproduce the characteristics of space, this activity will give you an idea of how dependent sound is on molecules of matter.

Materials

glass jar with a lid, about 236–250 mL (8 oz)
string, 10–12 cm long
tape
small bell
hot water

Procedure

1. Tape the string to the center of the inside of the lid.

2. Tie a small bell to the free end of the string.

3. Carefully screw the lid with the bell attached onto the jar. Make sure the bell does not touch the sides or bottom of the jar. If it does, adjust it by either cutting the string or moving the taped end of the string.

4. Gently shake the jar and listen to the bell.

5. Remove the lid with the bell attached and pour about 2 to 3 cm of hot tap water into the jar.

6. Allow the jar to stand for about 30 seconds and then replace the lid and bell. Be sure the bell does not touch the water.

7. Gently shake the jar again and listen to the bell.

8. Add more hot water and repeat steps 5, 6, and 7. Adjust the length of the string if necessary.

Observations and Conclusions

1. What did you observe in this activity?

2. How can you explain your observations?

3. What do your observations indicate about sound?

4. How does this activity relate to the lack of sound in space?

CUP-TO-CUP COMMUNICATION

When you were young, did you ever try to make a telephone out of paper cups and string? If so, your simple telephone may have been one of your first scientific endeavors! What you may not have realized, however, is that such a device utilizes the most basic principles of sound—the fact that sound is created by a vibration and travels as a disturbance that moves through materials such as air and string. In this activity you will get another chance to build a telephone, but this time you will be able to explain why it works.

Materials Needed

2 paper cups
frozen-dessert stick (or drinking straw or toothpick)
4 m cotton sewing thread (or dental floss)
sharpened pencil or scissors

Procedure

1. Use the sharpened end of a pencil or scissors to poke a small hole in the center of the bottom of each cup.
2. Pull the thread through the holes in the cups so that each cup has one end of the thread in it.
3. Break the frozen-dessert stick (or cut the straw or toothpick) in half. Tie each

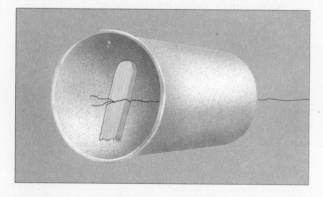

end of the thread to one of the halves so the thread cannot pull out of the hole in the cup.
4. Hold one cup in your hand and give the other cup to a classmate. Move apart from each other so that the thread is pulled tight. Be sure the thread does not touch anything.
5. Put your cup to your ear and have your classmate speak into the other cup. What do you hear? Now you talk and have your classmate listen. What does your classmate hear?
6. Switch roles again and this time try touching the thread while your classmate is talking. What happens?

Analysis and Conclusions

1. What is sound and how is it transmitted?
2. Use your definition of sound to explain how the paper-cup telephone works.
3. Can you explain why your observations changed when you touched the thread while your classmate spoke into the cup?
4. Compare the telephone you made to a real telephone.

Going Further

Would your results have been different if you had used aluminum or plastic containers instead of paper cups? If you had used knitting yarn or thin wire instead of thread (or dental floss)? Find out by repeating this activity using different materials. Remember, however, that to be truly scientific in your investigation, you can have only one variable!

JUST HANGING AROUND

You read that sound travels better through solids than it does through gases. Want to prove this idea for yourself? Try this activity.

Procedure

1. To begin, tie two strings (each about 30 cm long) to a metal hanger as shown in the accompanying diagram. (If you do not have a hanger, you can use two spoons.)

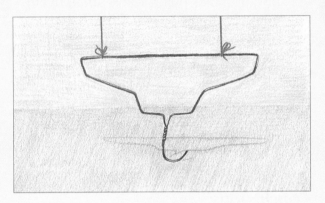

2. Hold the end of one string in one hand and the end of the other string in the other hand. Bump the hanger against a desk or other hard solid object. Listen to the sound produced.

3. Now wrap the end of each string around one of your index fingers.

4. Put your index fingers up against your ears and bump the hanger against the object again.

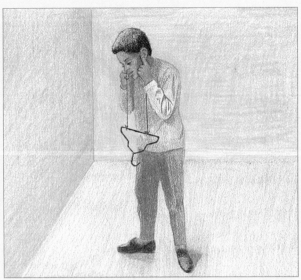

What I Proved

1. Describe the sound you heard the first time you bumped the hanger.

2. How does the first sound compare with the sound you heard with your fingers up against your ears?

3. What can you say about how sound travels through different mediums?

4. Can you now explain why people sometimes put drinking glasses up to doors or walls to hear through them?

Going Further

Try other solid materials in place of the hanger. How are they alike? How do they differ?

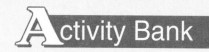

A SUNSATIONAL EXPERIMENT

Perhaps you have seen that old movie trick in which someone starts a fire using only a magnifying glass. Do you think it's a trick or can it really happen? Try the following activity to find out.

Materials

sheet of white paper
small glass bowl
magnifying glass

Procedure

1. Select a location that receives bright sunlight with no wind.

2. Crumple a sheet of paper and place it in a small glass bowl.

3. Hold a magnifying glass between the paper and the sun so a beam of light is focused on the paper.

4. Move the magnifying glass closer to and farther away from the paper. As you do so, notice how the point of light on the paper changes from a tiny bright circle to a hazy undefined shape. Adjust the distance between the paper and the magnifying glass to make the point of light very small and bright.

5. Once you have positioned the magnifying glass so that it forms a small bright circle, pull it back about 1 cm and watch the paper. **CAUTION:** *Under the right conditions, the paper may ignite.*

Observations

What did you see happening once you had positioned the magnifying glass correctly?

Conclusions

1. What do you think happened?

2. Why do you think this occurred?

3. How do your observations and conclusions compare with those of your classmates? How do you explain differences?

4. With this activity in mind, can you explain why a lawn will burn if it is watered during daytime sun?

MYSTERY MESSAGE

Have you ever imagined yourself to be an international spy using secret technology and writing messages in code—or better yet with disappearing ink? If so, you may be happy to discover that not all of your spy supplies are just in your imagination. In this activity, you will write a secret message that can be read only by someone who knows to use incandescent light to create the chemical reaction necessary to make the message visible.

Materials

toothpick
white paper, 1/2 sheet
lamp with light bulb
lemon juice (or lemon)

Procedure

1. Dip the toothpick into the lemon or lemon juice and use it as a pen to write a message on the white paper. Let the lemon juice dry. Observe the lemon juice as it dries.

2. Once the lemon juice has dried, hold the paper close to a lighted light bulb for several minutes.

Observations and Conclusions

1. What happens to the lemon juice as it dries?

2. What do you see when you hold the paper up to the light bulb?

3. A chemical reaction occurred when the paper was held up to the light. What are the two variables that could have caused the reaction?

4. The reaction you observed was an endothermic reaction. If you are not familiar with that term, look up the prefix *endo-* and the root *therm* to figure out what they mean. Now can you determine why the incandescent light bulb was used? Would a fluorescent bulb have worked as well?

5. Can you explain your observations in terms of chemical reactions?

THE STRAIGHT AND NARROW

You learned that although light travels in waves, light can be described by its straight-line paths. In this activity you will observe the straight path of a beam of light.

Materials

4 large index cards
flashlight or projector
metric ruler
scissors
8 chalkboard erasers (or small books or boxes)
thread or string (optional)

Procedure

1. Measure and mark the center of each index card. Then cut out a small circle (about 1 cm in diameter) in the center of each card.

2. In a room in which you can shut or dim the lights, space the cards about 30 cm apart. Hold each card upright so that the long side of the index card is on the floor or on a tabletop. Place chalk-board erasers on both sides of each card so that the cards stand upright without being held.

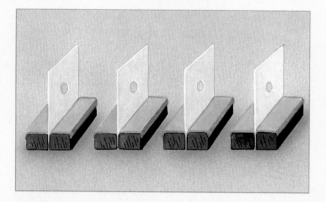

3. Make sure that the holes align. You may want to run a piece of string or thread through the holes in the cards and pull it tightly. This will help you to line up the holes.

4. Shut the lights so that the only light you see comes from the flashlight or projector. Now hold your light source so that it shines through the hole in one of the end cards. The light should not be able to shine around the first card. Have a classmate stand at the other end and observe.

5. Reverse roles with your classmate so that he or she holds the light source and you observe. What do you see?

6. Now move one of the cards about 3 cm to the side and repeat steps 4 and 5. Determine whether this changes what you observe.

Observations and Conclusions

1. What did you and your classmate observe in steps 4 and 5? In step 6?

2. What does this activity tell you about the path of light?

SPINNING WHEEL

As any painter knows, many of the colors you see are actually combinations of several colors. In this activity you will observe the results of combining different colors.

Materials

white posterboard	crayons or markers
compass	string, 1 m
metric ruler	India ink
scissors	several sheets of newspaper

Procedure

1. Use a compass to draw a circle with a diameter of 10 cm on white posterboard. Carefully cut out the circle using the scissors.

2. Divide the circle into three equal pie-shaped sections and color one section red, one section green, and the third section blue. Crayons or markers are the easiest coloring instruments to use.

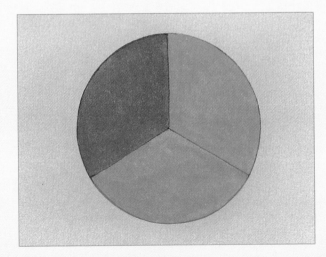

3. Use the pointed end of the compass or scissors to make two small holes on opposite sides of the center of the circle. They should be placed about 2 cm from each other. Your circle should look like a giant button.

4. Thread a string (about 1 m long) through the holes and tie the ends of the string together so that the thread forms a loop which passes through the holes. See the accompanying diagram.

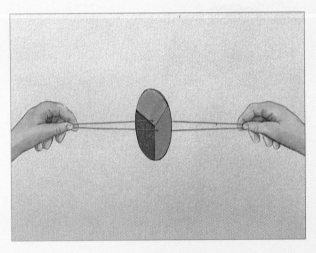

5. Center the circle on the thread, wind up the circle, and make it spin by alternately stretching and relaxing the string. It may take a little practice to keep it going. If you have trouble spinning the circle, you can try increasing its weight by doubling the thickness of the posterboard or pasting the circle onto cardboard.

The Next Step

Try the same thing with another circle with different colors and/or different numbers of sections. Predict each time what you expect to see.

6. Sometimes you see colors even when they are not there. Use the compass to draw another circle on the posterboard with a diameter of 10 cm. Cut out the circle with scissors.

(continued)

7. Cover your working surface with several layers of newspaper and place your white disk on top of the paper.

8. Use India ink and a paintbrush to paint your disk with one of the patterns shown.

9. When the ink is dry, again use the pointed end of the compass or scissors to make two small holes on opposite sides of the center. Thread string through the holes as you did before. Wind up the circle and make it spin. Observe how the circle looks.

Observations and Conclusions

1. What did you see when you spun the colored circle?

2. How can you explain your observations?

3. What did the inked circle look like when it spun? Can you hypothesize as to why you see what you do?

COLOR CRAZY

The color of an object is the color of light it reflects. A banana looks yellow because it reflects yellow light. However, in a beam of light that does not contain yellow, the banana looks black because there is no yellow light to reflect. In this activity you will have an opportunity to test this color theory for yourself.

Materials

shoe box
red and green cellophane (enough to cover the top of the shoe box)
scissors
flashlight
black construction paper
green object (such as a lime or green tomato)
yellow object (such as a banana or lemon)
red object (such as a red tomato or red playing card)

Procedure

1. Use the scissors to carefully cut a large rectangular hole in the lid of the shoe box. The hole should be large enough so that you can look through it and see anything placed in the shoebox.

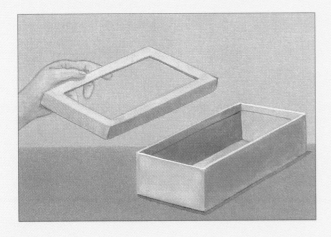

2. Cut a small hole in the center of one of the narrow ends of the shoe box.

3. Tape the green cellophane under the lid of the shoe box so that it entirely covers the hole you cut in step 1.

4. Line the inside of the shoe box with black construction paper.

5. Place the objects in the box and put the lid on.

6. In a darkened room, shine the flashlight into the shoebox through the small hole you cut out in the end. Record your observations.

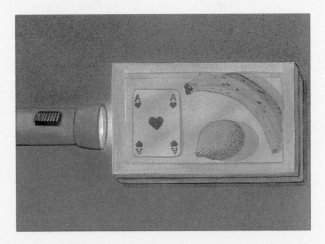

7. Replace the green cellophane with red cellophane and again shine the flashlight into the box. Again record your observations.

Observations and Conclusions

1. What did you see when the green cellophane was in place in step 6?

2. How can you explain your observations when looking through the green cellophane?

3. What did you observe when the red cellophane was in place in step 7?

(continued)

4. Were your observations different through the red cellophane? If so, why?

5. What is the purpose of the black construction paper?

6. What do you think would happen if you put a white object or an object with white on it in the box? Try it to see if you are correct.

7. Describe the role played by the cellophane in this activity. Draw diagrams to show what it does.

The Next Step

Repeat the activity using different objects inside the box and additional colors of cellophane.

Appendix A

The metric system of measurement is used by scientists throughout the world. It is based on units of ten. Each unit is ten times larger or ten times smaller than the next unit. The most commonly used units of the metric system are given below. After you have finished reading about the metric system, try to put it to use. How tall are you in metrics? What is your mass? What is your normal body temperature in degrees Celsius?

Commonly Used Metric Units

Length   The distance from one point to another

meter (m)   A meter is slightly longer than a yard.
1 meter = 1000 millimeters (mm)
1 meter = 100 centimeters (cm)
1000 meters = 1 kilometer (km)

Volume   The amount of space an object takes up

liter (L)   A liter is slightly more than a quart.
1 liter = 1000 milliliters (mL)

Mass   The amount of matter in an object

gram (g)   A gram has a mass equal to about one paper clip.

1000 grams = 1 kilogram (kg)

Temperature   The measure of hotness or coldness

degrees   0°C = freezing point of water
Celsius (°C)   100°C = boiling point of water

Metric–English Equivalents

2.54 centimeters (cm) = 1 inch (in.)
1 meter (m) = 39.37 inches (in.)
1 kilometer (km) = 0.62 miles (mi)
1 liter (L) = 1.06 quarts (qt)
250 milliliters (mL) = 1 cup (c)
1 kilogram (kg) = 2.2 pounds (lb)
28.3 grams (g) = 1 ounce (oz)
°C = 5/9 x (°F – 32)

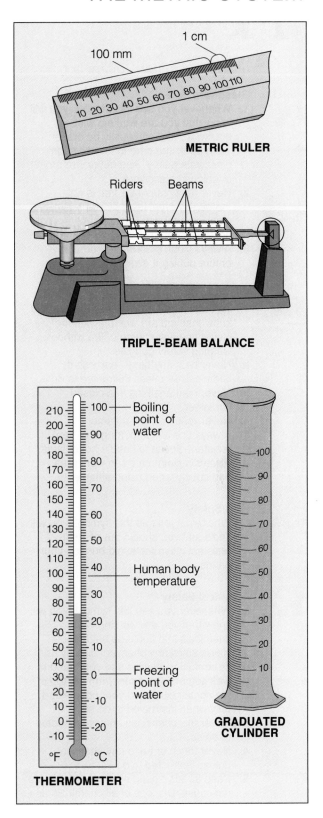

METRIC RULER

TRIPLE-BEAM BALANCE

THERMOMETER

GRADUATED CYLINDER

Glassware Safety

1. Whenever you see this symbol, you will know that you are working with glassware that can easily be broken. Take particular care to handle such glassware safely. And never use broken or chipped glassware.
2. Never heat glassware that is not thoroughly dry. Never pick up any glassware unless you are sure it is not hot. If it is hot, use heat-resistant gloves.
3. Always clean glassware thoroughly before putting it away.

Fire Safety

1. Whenever you see this symbol, you will know that you are working with fire. Never use any source of fire without wearing safety goggles.
2. Never heat anything—particularly chemicals—unless instructed to do so.
3. Never heat anything in a closed container.
4. Never reach across a flame.
5. Always use a clamp, tongs, or heat-resistant gloves to handle hot objects.
6. Always maintain a clean work area, particularly when using a flame.

Heat Safety

Whenever you see this symbol, you will know that you should put on heat-resistant gloves to avoid burning your hands.

Chemical Safety

1. Whenever you see this symbol, you will know that you are working with chemicals that could be hazardous.
2. Never smell any chemical directly from its container. Always use your hand to waft some of the odors from the top of the container toward your nose—and only when instructed to do so.
3. Never mix chemicals unless instructed to do so.
4. Never touch or taste any chemical unless instructed to do so.
5. Keep all lids closed when chemicals are not in use. Dispose of all chemicals as instructed by your teacher.

6. Immediately rinse with water any chemicals, particularly acids, that get on your skin and clothes. Then notify your teacher.

Eye and Face Safety

1. Whenever you see this symbol, you will know that you are performing an experiment in which you must take precautions to protect your eyes and face by wearing safety goggles.
2. When you are heating a test tube or bottle, always point it away from you and others. Chemicals can splash or boil out of a heated test tube.

Sharp Instrument Safety

1. Whenever you see this symbol, you will know that you are working with a sharp instrument.
2. Always use single-edged razors; double-edged razors are too dangerous.
3. Handle any sharp instrument with extreme care. Never cut any material toward you; always cut away from you.
4. Immediately notify your teacher if your skin is cut.

Electrical Safety

1. Whenever you see this symbol, you will know that you are using electricity in the laboratory.
2. Never use long extension cords to plug in any electrical device. Do not plug too many appliances into one socket or you may overload the socket and cause a fire.
3. Never touch an electrical appliance or outlet with wet hands.

Animal Safety

1. Whenever you see this symbol, you will know that you are working with live animals.
2. Do not cause pain, discomfort, or injury to an animal.
3. Follow your teacher's directions when handling animals. Wash your hands thoroughly after handling animals or their cages.

One of the first things a scientist learns is that working in the laboratory can be an exciting experience. But the laboratory can also be quite dangerous if proper safety rules are not followed at all times. To prepare yourself for a safe year in the laboratory, read over the following safety rules. Then read them a second time. Make sure you understand each rule. If you do not, ask your teacher to explain any rules you are unsure of.

Dress Code

1. Many materials in the laboratory can cause eye injury. To protect yourself from possible injury, wear safety goggles whenever you are working with chemicals, burners, or any substance that might get into your eyes. Never wear contact lenses in the laboratory.

2. Wear a laboratory apron or coat whenever you are working with chemicals or heated substances.

3. Tie back long hair to keep it away from any chemicals, burners and candles, or other laboratory equipment.

4. Remove or tie back any article of clothing or jewelry that can hang down and touch chemicals and flames.

General Safety Rules

5. Read all directions for an experiment several times. Follow the directions exactly as they are written. If you are in doubt about any part of the experiment, ask your teacher for assistance.

6. Never perform activities that are not authorized by your teacher. Obtain permission before "experimenting" on your own.

7. Never handle any equipment unless you have specific permission.

8. Take extreme care not to spill any material in the laboratory. If a spill occurs, immediately ask your teacher about the proper cleanup procedure. Never simply pour chemicals or other substances into the sink or trash container.

9. Never eat in the laboratory.

10. Wash your hands before and after each experiment.

First Aid

11. Immediately report all accidents, no matter how minor, to your teacher.

12. Learn what to do in case of specific accidents, such as getting acid in your eyes or on your skin. (Rinse acids from your body with lots of water.)

13. Become aware of the location of the first-aid kit. But your teacher should administer any required first aid due to injury. Or your teacher may send you to the school nurse or call a physician.

14. Know where and how to report an accident or fire. Find out the location of the fire extinguisher, phone, and fire alarm. Keep a list of important phone numbers—such as the fire department and the school nurse—near the phone. Immediately report any fires to your teacher.

Heating and Fire Safety

15. Again, never use a heat source, such as a candle or burner, without wearing safety goggles.

16. Never heat a chemical you are not instructed to heat. A chemical that is harmless when cool may be dangerous when heated.

17. Maintain a clean work area and keep all materials away from flames.

18. Never reach across a flame.

19. Make sure you know how to light a Bunsen burner. (Your teacher will demonstrate the proper procedure for lighting a burner.) If the flame leaps out of a burner toward you, immediately turn off the gas. Do not touch the burner. It may be hot. And never leave a lighted burner unattended!

20. When heating a test tube or bottle, always point it away from you and others. Chemicals can splash or boil out of a heated test tube.

21. Never heat a liquid in a closed container. The expanding gases produced may blow the container apart, injuring you or others.

22. Before picking up a container that has been heated, first hold the back of your hand near it. If you can feel the heat on the back of your hand, the container may be too hot to handle. Use a clamp or tongs when handling hot containers.

Using Chemicals Safely

23. Never mix chemicals for the "fun of it." You might produce a dangerous, possibly explosive substance.

24. Never touch, taste, or smell a chemical unless you are instructed by your teacher to do so. Many chemicals are poisonous. If you are instructed to note the fumes in an experiment, gently wave your hand over the opening of a container and direct the fumes toward your nose. Do not inhale the fumes directly from the container.

25. Use only those chemicals needed in the activity. Keep all lids closed when a chemical is not being used. Notify your teacher whenever chemicals are spilled.

26. Dispose of all chemicals as instructed by your teacher. To avoid contamination, never return chemicals to their original containers.

27. Be extra careful when working with acids or bases. Pour such chemicals over the sink, not over your workbench.

28. When diluting an acid, pour the acid into water. Never pour water into an acid.

29. Immediately rinse with water any acids that get on your skin or clothing. Then notify your teacher of any acid spill.

Using Glassware Safely

30. Never force glass tubing into a rubber stopper. A turning motion and lubricant will be helpful when inserting glass tubing into rubber stoppers or rubber tubing. Your teacher will demonstrate the proper way to insert glass tubing.

31. Never heat glassware that is not thoroughly dry. Use a wire screen to protect glassware from any flame.

32. Keep in mind that hot glassware will not appear hot. Never pick up glassware without first checking to see if it is hot. See #22.

33. If you are instructed to cut glass tubing, fire-polish the ends immediately to remove sharp edges.

34. Never use broken or chipped glassware. If glassware breaks, notify your teacher and dispose of the glassware in the proper trash container.

35. Never eat or drink from laboratory glassware. Thoroughly clean glassware before putting it away.

Using Sharp Instruments

36. Handle scalpels or razor blades with extreme care. Never cut material toward you; cut away from you.

37. Immediately notify your teacher if you cut your skin when working in the laboratory.

Animal Safety

38. No experiments that will cause pain, discomfort, or harm to mammals, birds, reptiles, fishes, and amphibians should be done in the classroom or at home.

39. Animals should be handled only if necessary. If an animal is excited or frightened, pregnant, feeding, or with its young, special handling is required.

40. Your teacher will instruct you as to how to handle each animal species that may be brought into the classroom.

41. Clean your hands thoroughly after handling animals or the cage containing animals.

End-of-Experiment Rules

42. After an experiment has been completed, clean up your work area and return all equipment to its proper place.

43. Wash your hands after every experiment.

44. Turn off all burners before leaving the laboratory. Check that the gas line leading to the burner is off as well.

Glossary

acoustics: science of sound and its interactions

amplitude: (AM-pluh-tood): greatest distance from rest to crest of a wave

atom: building block of matter

cochlea: part of the ear that contains hundreds of nerve cells attached to nerve fibers

concave lens: lens that is thicker at the ends than in the middle

concave mirror: mirror with a surface that curves inward

cone: nerve cell in the eye responsible for seeing color

convex lens: lens that is thicker in the middle than at the edges

convex mirror: mirror with a surface that curves outward

cornea: protective outer covering of the eye that refracts light

crest: high point of a wave

diffraction (dih-FRAK-shuhn): bending of waves around the edge of an obstacle

diffuse reflection: bouncing back of light from an uneven surface

Doppler effect: change in sound or light that occurs whenever there is motion between the source and its observer

eardrum: stretched membrane in the ear that vibrates at the same frequency as the sound waves that enter the ear

electromagnetic spectrum: arrangement of electromagnetic waves in order of wavelength and frequency

electromagnetic wave: wave that consists of electric and magnetic fields and does not require a medium to exist

farsightedness: condition in which the eyeball is too short, causing images to form behind the retina; corrected with a concave lens

fluorescent light: light produced by bombarding molecules of gas in a tube

focal point: location at which light rays reflected from a mirror meet

frequency (FREE-kwuhn-see): number of waves that pass a certain point in a given amount of time

fundamental tone: note produced at the lowest frequency at which a standing wave occurs

gamma ray: electromagnetic wave with the highest frequency and shortest wavelength in the electromagnetic spectrum

holography: technology that uses lasers to produce three-dimensional photographs

illuminated object: object that can be seen because it is lit up

incandescent light: light produced from heat

index of refraction: comparison of speed of light in air with speed of light in a certain material

infrared ray: electromagnetic wave in the frequency range just below visible light; felt as heat

infrasonic (ihn-fruh-SAHN-ihk): sound below the range of human hearing (20 Hz)

inner ear: liquid-filled portion of the ear; receives vibrations from the middle ear

intensity: amount of energy carried by a wave; indicated by the amplitude of a wave

interference: interaction of waves that occur at the same place at the same time

iris: colored area surrounding the pupil that controls the amount of light entering the eye

laser: *l*ight *a*mplification by *s*timulated *e*mission of *r*adiation; device that produces coherent light

lens: transparent material that refracts light

light ray: straight-line path of light

longitudinal (lahn-juh-TOOD-uhn-uhl) **wave:** wave in which the motion of the medium is parallel to the direction of the wave

luminous object: object that is capable of giving off its own light

mechanical wave: wave that disturbs a medium

medium: material through which a mechanical wave travels

microwave: high-frequency radio wave used primarily for communication

middle ear: part of the ear that receives vibrations from the eardrum and contains the hammer, anvil, and stirrup

modulation (mahj-uh-LAY-shuhn): variation; in particular, in the amplitude or frequency of an electromagnetic wave

nearsightedness: condition in which the eyeball is too long, causing images to form before the retina; corrected with a concave lens

neon light: cool light produced when electrons flow through a glass tube filled with gas

opaque: material that does not transmit light

optical fiber: thin tubes of glass used to transmit information as flashes of light

outer ear: part of the human ear that funnels sound waves into the ear

overtone: tone produced at frequencies higher than the fundamental at which a standing wave occurs

photoelectric effect: process by which light can be used to knock electrons out of a metal; can only be explained using the particle nature of light

photon: particle carrying energy that makes up light

pitch: property of sound that depends on frequency

plane mirror: mirror with a perfectly flat surface

polarized light: light in which all the waves are vibrating in the same direction

pupil: opening in the center of the eye through which light enters

radar: use of short-wavelength microwaves to locate objects and monitor speed

radio wave: electromagnetic waves with the longest wavelengths and lowest frequencies in the electromagnetic spectrum

reflecting telescope: telescope that uses a large mirror at its objective end

reflection (rih-FLEHK-shuhn): bouncing back of waves upon reaching another surface

refracting telescope: telescope consisting of two convex lenses at opposite ends of a long tube

refraction (rih-FRAK-shuhn): bending of waves due to a change in speed

regular reflection: bouncing back of light from a smooth, even surface

resonant frequency: frequency at which a standing wave occurs

retina: curved region in the rear of the eye on which images are formed

rod: nerve cell in the eye that is sensitive to light and dark

sonar: technique of using sound waves to measure distance; sound navigation and ranging

sound quality: blending of pitches to produce sound; timbre

standing wave: wave that does not appear to be moving; occurs at the natural frequency of the material

surface wave: wave that consists of a combination of transverse and longitudinal waves and occurs at the surface between two different mediums

timbre (TAM-ber): blending of pitches to produce sound; sound quality

total internal reflection: reflection that occurs if the angle of incidence for light is too great to be transmitted and is instead reflected back into its original medium

translucent: material that transmits light but no detail

transparent: material through which light is transmitted easily

transverse wave: wave in which the motion of the medium is at right angles to the direction of the wave

trough (TRAWF): low point of a wave

ultrasonic (uhl-truh-SAHN-ihk): sound above the range of human hearing (20,000 Hz)

ultraviolet ray: electromagnetic wave in the frequency region just above visible light

vibration: movement that follows the same path over and over again

visible light: colors of the spectrum that can be seen

wave: traveling disturbance that carries energy from one place to another

wavelength: distance between two consecutive similar points on a wave

wave speed: frequency of a wave times its wavelength

X-ray: electromagnetic wave in the frequency range just above ultraviolet rays

Index